AN APPROACH TO RHEOLOGY THROUGH MULTIVARIABLE THERMODYNAMICS

or

Inside the Thermodynamic Black Box

with
Addendum

by
Harry H. Hull

This book is dedicated to those who have a strong urge to master thermodynamics and who yet have an uncomfortable feeling that mastery is just around the corner.

Sponsored by the Society of Plastics Engineers

Harry H. Hull
care of Deeds Associates
1300 Benedum-Trees Building
Pittsburgh, PA 15222

Copyright
Harry H. Hull
1981 and 1982

This book is set in English Times type by Cold-Comp of Pittsburgh. It is printed and bound by Deeds Associates on Williamsburg Offset paper (a neutral sheet free of ground wood). The cover design of the paper back edition is by James Burke.

ISBN No. (Hardback with Addendum) 0-9606118-2-7
ISBN No. (Addendum) 0-9606118-3-5.

Foreword

There are many occasions when practitioners in plastics need to grasp some of the more basic concepts of their field. Rheology and thermodynamics represent the underpinnings for most correct interpretations of plastics processing. In this volume, *An Approach to Rheology through Multivariable Thermodynamics*, Harry Hull demonstrates that these difficult and somewhat abstract subjects can be presented in a lucid manner that informs and educates without burdening the reader with excess arithmetic baggage. In the continuing effort to provide its members and the plastics community as a whole with up-to-date technical source material, the Society of Plastics Engineers is pleased to sponsor and endorse this volume.

The Society of Plastics Engineers is dedicated to the promotion of scientific and engineering knowledge of plastics and to the initiation and continuation of educational activities for the plastics industry. To this end, SPE has sponsored books of this nature since 1956, with publication of at least one new technical volume annually in emerging segments of plastics not adequately covered in the literature. The Technical Volumes Committee is charged with the responsibility of surveying and determining the need for new topics, identifying authors and editors, recommending content and level of material, and, most importantly, reviewing final manuscripts for accuracy and relevance of technical material. On occasion, the Technical Volumes Committee recommends that a book be included in its growing paperback series. Harry Hull's book has met these criteria.

The Society prides itself on bringing to the plastics community high quality programs in meetings, seminars, educational courses, and publications. Its greatest resource is its 23,000 practicing plastics engineers, which make SPE the largest technical organization in plastics in the world.

It is with pleasure that SPE offers to the plastics community this volume by an outstanding researcher.

James L. Throne, Chairman
Technical Volumes Committee
Society of Plastics Engineers, Inc.
14 Fairfield Drive
Brookfield Center, CT 06805

THERMOWOCKY

'Twas quantig, and the vuscy graph
Unscrewed its curves with shrill delight;
The beaty ewes began to laugh
And slismal logs took flight.
"Beware the Entropy Beast, my son!
The mouth that sucks, the breath that cools!
Beware the Boltzmann bird, and shun
The chattering caloric ghouls!"
He took his dritting pen in hand:
Long time the beast of heat he sought —
Then rested he by the Plinck-Planck tree,
And stood awhile in thought.
And, as in thermal thought he stood,
The Entropy Beast, with eyes of flame,
Came slurping through the melting wood,
And siphoned as it came!
One, two! One, two! And through and through
The branksome pen went snicker-snack!
The head, the feet — and now the heat,
Released, came streaming back.
"And hast thou slain the Entropy Beast?
Accept this laurel for your head!
Oh frabjous day! Callooh! Callay!
The glarvish thief of heat is dead!"
'Twas quantig, and the vuscy graph
Unscrewed its curves with shrill delight;
The beaty ewes began to laugh
And the slismal logs took flight.

—*Alex Pelle*

(The above parody originally appeared in the *American Scientist* but was copied from the Chicago, American Chemical Society *Chemical Bulletin,* Oct., 1961)

It is hoped that the reader still has the sense of humor to appreciate the above parody on "Jabberwocky" from "Through the Looking Glass".

Preface

This book presents my personal viewpoints on thermodynamics. I consider them different than those held by most persons with knowledge of the area. At the same time these viewpoints are based on a close examination of the classical concepts of thermodynamics; however, these concepts are extended in unconventional ways to areas not covered in the texts. It is my hope and expectation that the viewpoints taken here will both show the reader new applications of thermodynamics and will reflect back on, and give the reader a better appreciation of, conventional equilibrium thermodynamics which uses only two independent variables of state.

My first solid introduction to thermodynamics was through the first edition of Lewis and Randall "Thermodynamics" in a course taught by Professor T. F. Young, a former student of Professor Lewis. I became interested in thermodynamics again through a very applied problem — how to describe the properties of printing ink. This is a problem in rheology, for the flow properties of ink at both low and high rates of shear are important in the printing process. As I saw the problem, a complete description of a body undergoing various rates of flow was analogous to the description of a gas at various states of compression, and hence both descriptions could be stated in terms of thermodynamics. However, the latter is a thermodynamics of the steady state rather than a thermodynamics of the equilibrium state.

The thermodynamics of the steady state is usually considered as one example of the thermodynamics of irreversible processes, Onsager (1931), Denbigh (1951), Prigogine (1955, 1962), Katchalsky (1965), Glansdorff (1971) etc. Most have been primarily concerned with the linear coupling of two or more irreversible processes or the flow and generation of entropy. I reviewed the pertinent literature in rheology, thermodynamics and statistical mechanics.† None of the approaches were satisfactory for my purposes so I tried various approaches of my own. I pursued one illusory approach for some time, which was finally discarded as invalid.

The approach which was finally developed and is described here is based on the extension of the methods of classical thermodynamics. Whereas classical thermodynamics is applied to equilibrium states and is stated in terms of two independent variables of state, the methods can be applied to the steady state and extended to three or more independent variables of state. It is also required and shown that the reversible and irreversible processes are separable, even though they may be coupled. One key concept is that classical thermodynamics is concerned with the properties of bodies (E, S, V, T, etc.) and the thermodynamics of the steady state is concerned with these same properties (as well as additional ones).

The subtitle of this book "Inside the Thermodynamic Black Box" has at least three interpretations, and all are meant to be valid. First, it covers

---

†Lavenda (1978) reviews and gives a critical evaluation of the current approaches to irreversible thermodynamics.

the algebra of thermodynamics extending it to three independent variables of state. Second, it considers the thermodynamics as the study of various mechanical systems including their relationships to heat (and other forms of energy). Third, it interprets thermodynamics in terms of the action and properties of the molecules and atoms in that black box.

The algebra of thermodynamics presented here is certainly within the classical tradition. Its extension to three independent variables of state is not new, but in my opinion its utility and significance have not been appreciated.

The models of thermodynamic systems described here are certainly within the older traditions of thermodynamics. The emphasis placed on them is in part due to my belief that the algebra must apply to concrete thermodynamic models, and that a visualization of these models demystifies much of thermodynamics. Thermodynamics should not be a rote learning and rote application of equations. The visualization of system models to which the equations apply avoids the rote learning and should help in the prevention of errors in the application of thermodynamics.

The interpretation of thermodynamics in terms of the actions and properties of molecules is a form of statistical mechanics. However, most of the discussion here is in terms of qualititative models, so that a mental picture of the molecular actions is attained. Such a mental picture of the actions of the molecules aids in the understanding of thermodynamics.

This qualitative approach to statistical mechanics also avoids the partition function on which much of modern statistical mechanics is based. This is avoided in part because its use for classical systems involves the perfect gas assumption (and hence is defective for real systems) and in part because it does not fit well with the visualization of models. Quantitative approaches to the statistical mechanics of long chain molecules are available in the literature with the work of Rouse (Ferry, 1970) and Flory (1953, 1969).

The one quantitative approach to statistical mechanics in this book is in the chapter on entropy. This whole chapter came about because I had not found in the literature any (satisfactory) proof that the statistical definition of entropy and the thermal definition of entropy defined identical and not just related quantities. This problem goes back to the time of Boltzman (1964), (Brush, 1976) and involves the ergodic and quasi-ergodic hypotheses. In my opinion most of the discussion in this area tends only to obscure the problem. The correct statement is that time averages are equal to the probability in the statistical definition of entropy ($S = k \ln W$), and that no valid method for the calculation of the times averages had been developed.† In Chapter IV a method is proposed for the calculation of time averages for the simplest real case — heavy monatomic gases. The method is confirmed by showing that answers are obtained which agree with those obtained by the classical methods derived from the thermal devinition of entropy. All of my questions about the identity of the two definitions of en-

†This is also the viewpoint of Kinchin (1949) and some others.

tropy have not been answered; however, these calculations demonstrate that they are equal in two cases of simple real substances.

The chapter on entropy is somewhat of a diversion from the rest of the book. To me it was undertaken primarily as an exercise to develop an understanding of entropy. It has been included here with the hope that it will do the same for the reader. However, it does propose a novel solution to a previously unsolved problem.

The first part of this book contains a review of some of the classical methods of thermodynamics. This review is necessary; for, a certain viewpoint of these methods is required before they can be shown to be valid for the steady states of rheological systems.

There are many traps in the application of thermodynamics to the steady state. However, these are avoided by the help of (a) careful definitions of the appropriate equations of state, (b) careful notation of *all* the variables which are held constant in the partial differential equations, (c) the addition of a new notation which designates the variables which are held constant in the Δ functions, (d) the use of extended thermodynamic functions which contain the additional variables of state, and (e) a careful description of the models to which the equations apply.

Although this presentation of thermodynamics is substantially different from the conventional, I believe that a careful reading of the whole book in the order of presentation will be understandable to a reader with a background in physical chemistry and who has had an introduction to thermodynamics. I have tried to make a simple and concrete presentation and avoid obscure and complicated statements. However, those who are trained in the concept that the classical methods of thermodynamics only apply to "equilibrium states" will have to adjust their concepts.

To those who (I am sure justifiably) approach this book with skepticism I plead that they read it in the order of presentation — without skipping around. There is a certain logic which requires the sequence of presentation, and which I hope will be clear to such a reader.

I wish to thank the Society of Plastics Engineers for their encouragement and cooperation in the writing of this book. This applies in particular to Dr. Thomas Haas and Dr. James L. Throne, the chairmen of the Technical Volumes Committee during its writing and publication.

I also wish to thank the various persons who have read the manuscript at various stages and submitted their comments to me. They have been helpful.

Harry H. Hull
September 1980

Permission to use figures or diagrams has been granted by the following and is gratefully acknowledged.

Academic Press, Figure IX-6, from Rumscheid, F.D., *J. Col. Sci.*, *16*, 238, (1961).

Annual Reviews, Inc., Figure IX-10 from Cox, R.G. and Mason, S.G., from *Ann. Rev. Fluid Mech.*, *3*, 291, (1971).

Pergamon Press Inc., Figure VII-12, from Wriedt, H.A. and Oriani, R.A., *Acta. Met.*, *18*, 753, (1970).

Canadian Journal of Chemistry, Figure IX-11, Okagawa, A. and Mason, S.G., *Can. J. Chem.*, *53*, 268, (1975).

Table of Contents

Foreword		III
Preface		V
Chapter I	A Viewpoint on Thermodynamics	1
Chapter II	The Equations of State	5
Chapter III	The First Law and the Thermodynamic Variables Pressure, Temperature, Volume and Internal Energy	13
Chapter IV	The Two Definitions of Entropy and the Second Law of Thermodynamics	21
Chapter V	The Thermodynamic Functions	31
Chapter VI	Models of Systems in Balance	39
Chapter VII	The Thermodynamics of Elastic Deformation	49
Chapter VIII	The Separability of the Reversible and Irreversible in Thermodynamic Processes	75
Chapter IX	The Thermodynamics of Viscoelastic Fluids	91
Chapter X	The Thermodynamics of Systems with Gradients	119
Chapter XI	The Relationship between the Thermodynamics of Rheology and Chemical Thermodynamics	129
Chapter XII	Recapitulations and Speculations	139
Appendix I	List of Principal Symbols	145
Appendix II	References	149
Author Index		155
Subject Index		157
Addendum		A-1

Chapter I
A Viewpoint on Thermodynamics

The number and variety of books on thermodynamics is amazing. Each author presents his own viewpoint which is different in one or more aspects from the others. One particular viewpoint is not necessarily wrong and the other right.

Thermodynamics is like a real object in three dimensional space. The shape of the object is such that its appearance varies radically from one viewpoint to another.

Thermodynamics is also like a country with a network of roads. One may enter the country at a number of different entrances and the initial impression of that country depends on the characteristics of that country at the initial entrance and on the roads which one first travels. It may take some time before the country is well explored by an individual, and two individuals who have explored that country may have different opinions of that country — depending on their particular experiences and on their opinions before they enter the country. Of course if two individuals have a common guide their opinions are likely to be similar. A guide helps initially, but to know a country one must explore it on his own.

Thermodynamics is also a means for describing the behavior of a class of black boxes. On the outside of the boxes are a number of levers which have indicators of their position. These are lever-indicators. There are also a number of gages which are only indicators. There are a tremendous number of these levers and gages and they are not always in a prominent position or easy to operate or read. In order to use the black box to its fullest advantage one must know what levers and gages to look for and how to look for them.

Some of the levers are connected together on the outside in ways that are very apparent; however, as the box originally comes the levers flop around loosely for there is nothing in the box. There is however, a door on the box through which materials can be inserted. These boxes come in various sizes and one selects the size which is the most convenient for his purposes. Very small boxes take only a few molecules, while there is no upper limit on the size of the boxes. The most convenient size of box takes one gram molecular weight of material.

Once a material has been inserted in the box the levers† no longer flop around loosely; however, they can be moved by the operator. The material in the box acts as though it were gears, cams or internal arms connecting the

---

†The lever-indicators from here on will be referred to as simply levers.

outside levers. The simplest type of box contains only one phase (i.e., only a gas, or a liquid, or a solid). For such a box two different levers can be moved simultaneously but not three or more, as the positions of any two levers determine the positions of the rest. The most prominent of the levers are labeled P, T, and V. In a somewhat less prominent position are E and S. The connections between the levers on the outside of the box confine their movement so they obey the laws of thermodynamics. The thermodynamic functions are indicated by an analog computer whose values are computed only from the positions of the outside levers.

There are all sorts of gages on the black box which are read when a button beside them is pressed. These have labels such as optical index of refraction, electrical conductivity, thermal conductivity, specific heat, etc. Some of these gages are covered and represent properties of matter which are presently unknown.

The black box will also accept models of materials which may or may not exist. The best known such model is the ideal gas. These models tend to lack some properties (as indicated by the gages) that all real substances have — for example viscosity, optical index of refraction, or even thermal conductivity. Sometimes these properties can be incorporated by modifying the model.††

The single phase black box has a number of catches on the side. Each of these catches when released will in turn release two additional levers; however, only one of this pair can be considered independent as they are linked together by the properties of the material inside the box. Such pairs of levers represent additional variables of state such as: stress and strain, surface tension and surface area, magnetic field and magnetic induction, etc. When one catch is released, three levers can be moved simultaneously and independently as there are then three independent variables of state. When two catches are released, four levers may be moved simultaneously and independently.

The black box also has provisions for the attachments of various devices on the outside. Examples of these attachments are devices which produce electrical or magnetic fields. These devices can operate when the black box is empty, for electrical and magnetic fields can exist in a vacuum. Devices to show this are on the levers rather than on the black box. When a material is in the black box the levers and gages indicate separately the electric and magnetic response of the material and the response of the empty black box.

There are many dials and levers on the black box and we must find by experiment which respond and which do not with various materials in the box. For example, tension or compression on some materials creates an electrical chrge (piezoelectric materials), and some develop a charge when their temperature is changed (pyroelectric materials).

---

††Assuming a molecular weight for the ideal gas will permit calculation of viscosity and thermal conductivity.

Some black boxes can hold materials in the steady state. Such boxes are provided with a means for energy and material to flow into and out from the box in such a way that both the material in the box and its energy remain constant. Viscoelastic fluids fit into these steady state boxes. The early chapters in this book are prelude to the demonstration that they do fit in and also demonstrate which of the levers and gages operate when it contains such a fluid.

Chapter II
The Equations of State

One of the concepts on which classical thermodynamics is based is that of the equations of state. However, a complete discussion of their derivation, uses, limitations, and significance is not included in the textbooks or literature on thermodynamics. The omission is probably made in part because the equations of state are taken to be self evident. A detailed review and discussion illuminates this neglected area in the foundations of thermodynamics.

The concept of the equation of state is familiar but usually implied rather than stated or taught as a principle. The gas laws, the tables of density of water at various temperatures and pressures, tables of the viscosity at various temperatures, tables of the strength of materials, and steady state descriptions all imply equations of state. These concepts need to be brought together, summarized, and broadened, so they may have their utmost utility in both classical and irreversible thermodynamics.

It is best to begin by reviewing the equations of state in their simplest form and then add to them step by step until the equations of state are expressed in their broadest form.

(a) Two examples of equations of state are:

$$V = f(P, T) \tag{1}$$

$$E = f(P, T) \tag{2}$$

Equation (1) states that volume per mole (of a particular substance which may be either a gas, liquid, or solid) is a function of pressure and temperature. Equation (2) states that the internal energy per mole is also a function of pressure and temperature. These equations of state derive their validity from, and at the same time are an expression of, laboratory experience. This experience is that once P and T are known and V and E have been established for those particular values of P and T, V and E have been established for that gas, liquid, or solid for any time in the past, present, or future.

Pressure and temperature are the independent variables of state, and volume V and internal energy E are the dependent variables of state.

Specific equations of state such as the Van der Walls or the Berthelot are not pertinent to this discussion.

(b) Equation (1) could have been written,

$$f(P, T, V) = 0 \tag{3}$$

$$\text{or, } T = f(P, V) \tag{4}$$

however, neither is the preferred form as neither is necessarily single valued. An example in which (3) and (4) are not single valued is water. Near its freezing point water can have the same volume at two different temperatures. The properties of those variables of state which are most desirable to select as the independent variables will be discussed later.

(c) A more general way of writing the equation of state is:

$$B = f(P, T) \tag{5}$$

Where B symbolizes the volume per mole, the internal energy per mole, and all the other dependent variables of state. These dependent variables include the thermodynamic functions such as the enthalpy, the Gibbs function, the specific heat, and such properties as the optical index of refraction and the viscosity. The thermodynamic functions will be discussed in detail in Chapter V.

(d) Although it is not written, the equation of state implies that the composition is constant. A more precise method of writing (5) is,

$$B = f(P, T)_{m_1, m_2 \cdots m_n} \tag{6}$$

where m_n represents the weight of component n, and its use as a subscript indicates that it is constant. The quantity B is written in large capitals instead of small capitals for it is not necessarily expressed as units per mole but as units per designated amounts of specific components.

The various components may also be included as independent variables and the equation written:

$$B = f(P, T, m_1, m_2 \ldots m_n) \tag{7}$$

Sometimes it is convenient to express the quantity of materials in terms of mole fractions (x_i). In this case the equation of state becomes:

$$B = f(P, T, x_1, x_2 \ldots x_{n-1}) \tag{8}$$

(e) These equations apply to a body of one phase. The phase may be a solid, a liquid, or a gas but not a mixture of two phases (except in a very special and specific case to be mentioned later). For example there is one equation of state for water vapor, one for liquid water, and one each for each of the crystalline forms of ice.

(f) These equations of state apply to phases at equilibrium; however, they also apply to certain phases which are in an unstable state, for example: superheated or supercooled water, supercooled solutions, ammonium nitrate (an explosive), mixtures of hydrogen and oxygen. These materials are in a metastable state or (when chemical reactions are involved) its equivalent "false" equilibrium". They can change spontaneously; or change when subjected to disturbances such as, mechanical shock, a dust speck, or a catalyst.

These equations of state also apply to bodies which are known to be changing very slowly — such as materials which undergo radioactive decay, or rubber which ages. They then of course apply only within the time limit for which the amount of slow change is negligible.

(g) One of the uses of the equations of state is that they are the source of an important series of partial differential equations (Lewis and Randall, p. 25 [1961]). For example from (1) above we obtain,

$$(\partial V/\partial T)_P = -(\partial V/\partial P)_T (\partial P/\partial T)_V \qquad (9)$$

and from (2) above,

$$(\partial E/\partial T)_P = -(\partial E/\partial P)_T (\partial P/\partial T)_E \qquad (10)$$

When the above equations are combined with other partial differential equations obtained by other methods new and useful thermodynamic relationships are obtained. If B is used as a generalized dependent variable the equation becomes,

$$(\partial B/\partial T)_P = -(\partial B/\partial P)_T (\partial P/\partial T)_B \qquad (11)$$

or,

$$(\partial B/\partial T)_{Pm} = -(\partial B/\partial P)_{Tm} (\partial P/\partial T)_{Bm} \qquad (11a)$$

where the subscript m indicates that the mass and composition are constant.

(h) Pressure, temperature, and composition as given in equation (7) are not the only independent variables of state. Consider a piece of rubber as a thermodynamic body. If the rubber is stretched, work is done on it, and its molecules will be oriented. If kept at constant temperature while it is being stretched, heat will be transferred to its surroundings. The properties of the rubber and hence the state of the rubber have been changed. There are equations of state which contain more than two independent variables of state.

Thermodynamic variables generally come in pairs.† One of these variables is intensive and force-like. The second of the pair is its conjugate and it is extensive. The product of the two is work or energy. This product is used to form the thermodynamic functions which will be discussed in Chapter V.

Pressure is an example of an intensive variable. Its conjugate is volume.

Temperature is another intensive variable. Its conjugate is entropy.

Stress is an intensive variable whose conjugate is strain.

The extensive variables of state can be converted to intensive variables by expressing them in terms of units per unit weight. Such quantities are molal volume v, and molal entropy s. Strain can also be expressed as a ratio which is intensive. (Such a ratio must be multiplied by volume to obtain the correct units of energy when multiplied by stress). Extensive variables when converted to an intensive form do not have the force-like property.

As a general rule it is preferred to consider the force-like intensive variables as the independent variables of state, and their conjugates as the dependent variables.

---

†The internal energy E is an exception.

If the intensive variables of state are uniform throughout the thermodynamic body the equation of state can be written,

$$B = f(P, T, X_3 \ldots X_i \ldots X_s, m_1 \ldots m_n) \tag{12}$$

where there are s independent variables of state and n components. The B represents all the dependent variables of state including the conjugates of the force-like intensive variables.

If all the "X_i"s in (12) are held constant at zero value the equation reverts to (7). This justifies the two independent variables of state of conventional thermodynamics.

Other variables of state include those of the magnetic, electric, gravitational, and centrifugal fields. Surface tension with its conjugate surface area are also variables of state. Thermodynamic systems involving some of these variables (though not stress) are treated in some of the standard thermodynamic texts. (Lewis and Randall [1961], Guggenheim [1967]).

(i) The bodies to which the above equations of state have been applied were assumed to be in either thermodynamic equilibrium or in the metastable state. Equations of state also exist for bodies to which energy is constantly being added and taken away at the same rate. These bodies are in the steady state.

For example, consider a fluid body undergoing shear but held at a constant temperature. Work is being done on the fluid, the work is converted to heat, and the heat is absorbed by a constant temperature bath. If the fluid is not shear sensitive (such as water) the properties of the fluid will not change appreciably when undergoing shear. If the fluid is shear sensitive, the viscosity of the fluid will change. It may also become birefringent, and/or it may develop a recoverable elastic displacement. All of these are properties of state, so equation (12) applies. If the stress in shear is selected as an independent variable of state its conjugate is the recoverable elastic displacement. Their product is work just as with the stress and strain of elastic solids. Since the force-like variables may or may not make appreciable changes in the properties of a body it is desirable to keep the force-like variables as the independent variables of state, for then the equations of state are single valued. Note that equation (12) is valid in this case whether or not the fluid body is shear sensitive. When elastic displacement is small the elastic displaement is not a useful variable of state. If viscosity is used as an independent variable of state, the equation of state will not be single valued for Newtonian fluids.

Another steady state to which an equation of state can be applied is the steady state in the presence of a thermal gradient. Consider a polymer solution interposed between two constant temperature sources T_1 and T_2. The flow of heat from the higher temperature to the lower will gradually reach a steady state and may be accompanied by the development of birefringence or a concentration gradient. The latter is the Soret effect (Denbigh p 19, [1951]; Tykodi chapter 8, [1967]). It is evident that the state of the polymer solution has been changed by the thermal gradient. However, specific

THE EQUATIONS OF STATE

numerical values of the equation of state (12) can only be applied to infinitesimal lengths along the thermal gradient for the temperature and other variables change along that gradient.

(j) The partial differential equations mentioned in (g) also apply when other independent variables of state are present. One must, however, be careful to note that with three independent variables, two variables must be kept constant for each partial differential. This should be noted by the use of both variables as subscripts. For example, if P is kept constant and X_3 is added as an independent variable of state, an equation analogous to (11) is obtained,

$$(\partial B/\partial T)_{PX_3} = - (\partial B/\partial X_3)_{PT} (\partial X_3/\partial T)_{BP} \qquad (13)$$

This may be expanded by substitution to,

$$(\partial B/\partial T)_{PX_3} = (\partial B/\partial P)_{TX_3} (\partial P/\partial X_3)_{BT} (\partial X_3/\partial T)_{BP} \qquad (14)$$

(k) An important use of the equations of state is that the phase rule can be developed from them, for the equations of state define a system of simultaneous equations. Consider a mixture of n components in p phases at equilibrium with P and T as the two independent variables of state. There are $n-1$ variables of composition for each phase. The partial vapor pressure of each component (or its fugacity) can be taken as the "B" or the dependent variables, and these must be equal for each component throughout the system. There are then n equations of state for each phase or np simultaneous equations.

The total number of variables in the equations of state are,

P and T	2
A vapor pressure for each component	n
$n-1$ variables of composition for each phase	$p(n-1)$
Total number of variables	$n - p + 2 + np$

If the total number of variables equal the number of equations, i.e.,

$$np = n - p + 2 + np \qquad \text{or}$$
$$n - p + 2 = 0$$

there will be only one solution to the set of simultaneous equations, Hence, the system could exist at only one point, i.e., at a particular pressure, temperature, and particular compositions for each phase. If the total number of variables is greater than the number of equations, i.e.,

$$n - p + 2 > 0$$

the simultaneous equations will have more than one solution. The number of variables which must be fixed before the simultaneous equations must be fixed is known as the degrees of freedom. This is expressed as the well known phase rule,

$$r = n - p + 2 \qquad (15)$$

where r is the degrees of freedom of the system.

Zernike (p. 6, 1955) showed the development of the phase rule by the above method.

If there are more than two independent variables of state, one more degree of freedom is added for each new independent variable of state, and the phase rule becomes,

$$r = n - p + s \qquad (16)$$

where s is the number of independent variables of state. Since the equations of state are valid for steady states as well for equilibrium, the phase rule is also valid for steady states (and metastable states).

The phase rule applies to liquid crystals and Adams (1971) has commented on the effect of other variables on their behavior: "In addition to a wide variety of thermally induced transitions there are an abundance of electric field, magnetic field, shear, etc. induced phase changes."

The phase rule provides another illustration for the desirability of using the force-like variables as the independent variables of state. If they are, the equation of state will be single valued and the phase rule (equation 16) is unique. If some other variable is selected as the independent variable of state their solution is not necessarily unique and such equations cannot be used to develop the phase rule. The best known example of equations of state which are not single valued are those for water near its freezing point. The equation of state which uses P and T as the independent variables and V as the dependent variable of state is single valued. If P and V are selected as the independent variables the equation of state is not single valued for there are two temperatures at which P and V are the same.

(m) There is an occasional mention in the literature of "internal variables of state". This implies there are variables of state analogous to P and T which either cannot be identified or whose level cannot be measured. If there are "internal variables of state" which affect the properties of a body, this is equivalent to the body being out of equilibrium in terms of external force-like variables of state. Examples of bodies which may be thought of as having "internal variables of state" are those with a memory, bodies which show hysteresis, and bodies with incomplete chemical reactions. These are traditionally considered to be out of equilibrium (or in terms to be defined later to be out of thermodynamic balance).

However, it is sometimes possible to express the state of a body which is out of thermodynamic balance in terms of *extent of reaction* of a physical-chemical change (Guggenheim, p 62, 1967). It is not necessary that this extent of reaction be completely controllable provided that its value can in principle be measurable. It can refer to a frozen state, or to one at a particular instant of time when rapid changes are occurring. In the literature on chemical thermodynamics it is designated as ξ.

(n) It is interesting to note Callen's (1961) approach to equations of state for multiple variable systems, which he designates as general systems. His postulate I (p. 192) is an hypothesis of the existence of the equation of state (in the terminology of this paper).

$$B = f(E, V, Y_1 \ldots Y_s) \tag{17}$$

where the "Y"s are the extensive parameters of state. He assumes that the intensive parameters are functions of the extensive parameters, and develops a thermodynamics of elastic solids on this basis. This is valid though it varies from the approach used here where the intensive variables of state are the preferred independent variables.

(o) In summary the equations of state underlie all of thermodynamics, and as such are as basic to it as the first and second laws of thermodynamics. The equations presented here apply only to bodies which can be described by equations of state. For the equations of state to apply, a body must meet one of the four conditions: (1) be at equilibrium, (2) be in the metastable state or in a "false equilibrium", (3) be in the steady state, or, (4) be in a state that the extent of reaction of a physical-chemical change is measurable in principle.

Chapter III

The First Law and the Thermodynamic Variables, Pressure, Temperature, Volume and Internal Energy

This Chapter is a short discussion of selected material generally found in the literature on physical chemistry, so few references or detailed proofs will be given. It is presented here partly to make this book more complete and partly to present a particular viewpoint in preparation for the following chapters.

The First Law

The first law of thermodynamics is that energy is neither created or destroyed though it may change in form — exist as work, or heat, or chemical energy, etc. The law is repeatedly applied and is such an intrinsic part of thermodynamics that it is usually applied without citation of the law. The concept of the first law underlies most thermodynamic equations.

Internal Energy

The internal energy E can be thought of as a quantity devised to fulfill the first law so that the heat added to a body is equal to the work done by the body plus the energy retained in the body,

$$dq = dE + dW \tag{1}$$

Although the above is valid, it is better to take a more positive approach and visualize the internal energy as an actual energy made of a number of components — the translational, rotational, and vibrational kinetic energies of the molecules along with their associated potential energies. At higher temperatures the internal energy includes energy changes from the changes in electron orbits.

The Ideal Monatomic Gas

The hypothetical ideal monatomic gas follows the equation,

$$PV = NkT \tag{2}*$$

or its equivalent equation expressed on a molal basis

$$PV = N_o kT \tag{2a}*$$

---

*Equations valid only for ideal gases or ideal rubbers are marked with *.

All its internal energy hypothetically exists as translational kinetic energy of the molecules. The molecules of this gas have mass but no volume, and they do not attract or repel each other. The internal energy is a function of temperature alone, hence,

$$(\partial E/\partial V)_T = (\partial E/\partial V)_T = 0 \qquad (3)*$$

The internal energy is proportional to the absolute temperature, hence, the specific heat at constant volume is,

$$Cv = (\partial E/\partial T)_V = (\partial E/\partial T)_P = (3/2) Nk \qquad (4)*$$

or on a molal basis

$$Cv = (\partial E/\partial T)_V = (\partial E/\partial T)_P = (3/2) N_o k \qquad (4a)*$$

The specific heat at constant pressure includes the work of expansion,

$$Cp = Cv + P(\partial V/\partial T)_P = (5/2) Nk \qquad (5)*$$

The translational motion has three degrees of freedom with one third of the total specific heat at constant volume associated with each degree of freedom.

The distribution of molecular velocities (v) for an ideal monatomic gas was determined by Maxwell to be,

$$dN/N_o = 4\pi(m/2\pi kT)^{3/2} \exp(-mv^2/2kT) v^2 dv \qquad (6)*$$

In developing this expression Maxwell made two assumptions: (1) The probability of the velocity of a molecule parallel to one axis is independent of the probability of the velocity parallel to the other two orthogonal axes, and (2) The velocity is independent of the position of the molecule (with respect to the other molecules). It is the latter assumption which limits its validity to the ideal gas (and to real gases whose behavior approximates the ideal gas). This latter assumption will be discussed further in Chapter IV.

Most gases follow the ideal gas law at low pressures and moderate temperatures. The gases with low liquefaction temperatures (such as helium and hydrogen) follow the ideal gas law over a greater range of pressures and temperatures.

The ideal gas is both a blessing and a curse to thermodynamics. It is a useful approximation for many purposes. On the other hand approximations based on ideal gas assumptions are not always clearly identified so large errors can creep into calculations without the realization that the ideal gas approximation is involved.

Pressure

The pressure exerted by a gas is the force per unit area exerted by the molecules bounding against the walls. When the area over which the pressure is exerted is very small this force will vary rapidly, depending on the number and velocity of the molecules hitting the wall at any instant. The time average of the pressure exerted over the very small area is equal to the expected value of the pressure written as $<p>$. It is also equal to the average pressure over a large area measured at an instant of time. Actual pressure measuring devices do not respond to the rapid changes in pressure associated with the motion of the molecules and they measure pressure over a relatively large area. The pressures recorded by instruments and listed in thermodynamic tables can be regarded as either time averages or area averages.

Temperature

Temperature is a variable which has a personal meaning. Our bodies are sensitive to small variations in temperatures and we are familiar with fire and the use of higher temperatures to cook food and with lower temperatures to preserve food. The concept of temperature in thermodynamics is more subtle and we must probe into the physical meaning of temperature.

A qualitative concept of temperature is given by the observation that heat always flows from a higher temperature to a lower temperature and by the "zeroth" law of thermodynamics — if two bodies have the same temperature as a third, their temperatures are equal.

In order to make temperature quantitative a temperature scale must be established. This has two aspects, (1) the establishment of two reference temperatures, and (2) the establishment of a scale between and beyond the two reference points. The two reference points have traditionally been the boiling and freezing points of water but now are the triple point of water and absolute zero. The temperature scale which has been accepted between and beyond the reference points can be established by three different methods each of which results in the same temperature scale;

(a) The ideal gas temperature scale in which the volume of an ideal gas held at constant pressure is proportional to the temperature (or proportional to the pressure of an ideal gas held at constant volume). In terms of laboratory measurements using real gases, temperature is proportional to the product PV as pressure approaches zero.

(b) Temperature is the factor θ the inverse of which is a multiplier which makes the heat absorbed δq a perfect differential and q/θ (entropy) a measure of energy degradation. Note that Lord Kelvin called θ the thermodynamic temperature.

(c) Temperature is proportional to the translational kinetic energy of an ideal monatomic gas. This definition has been stated before in equation (III-4) for the specific heat of an ideal monatomic gas. The internal energy of all ideal monatomic gases is independent of the weight of the molecules, is independent of the volume in which they are contained, and depends only on temperature.

Consider that two ideal monatomic gases of different molecular weight are placed in a container with a thin impermeable membrane between them. This membrane has imaginary properties such that the molecules of the two gases can collide in the region of the membrane but cannot pass through the membrane. The gases cannot mix. By the process of colliding the average translational kinetic energy of the molecules of each gas will become equal. The equilibrium is established between the kinetic energies of the two gases. If one of the ideal gases is replaced by a real gas which also cannot pass through the membrane and one which can collide with the other ideal gas molecules, equilibrium will be established at which the average translational kinetic energies of the two gases become equal. The relative magnitude of the intromolecular potential energies as well as the presence of kinetic energies which are not translational do not affect the balance between the kinetic energies.

The concept of temperature involves the concept of absolute zero — where the kinetic energy of the molecules of the ideal monatomic gas is zero. With real substances the absolute zero is also reached when there is no translational kinetic energy in the molecules or atoms. Any other motion which could exchange energy with the translational motion would also have to be zero. Note that the intermolecular potential energy does not have to be at a minimum to attain absolute zero in temperature.

The actual magnitude of a temperature degree is unimportant and can be arbitrarily defined. For example, the Celsius and Fahrenheit degrees (as used in the Kelvin and Rankine absolute temperature scales) can be used with equal facility; however, changing the temperature scale involves changing some thermodynamic constants.

It is interesting to note that certain constants could be simplified by redefining the degree. For example, if the degree were defined as the change in temperature of one mole of an ideal gas on the addition of 1/8.3143 joules of energy the universal gas constant R_o would be 1.0 in units of joules per mol per new degree. If a degree were defined as the change in temperature when 1.3806×10^{-16} ergs of energy were added per molecule of an ideal monatomic gas the value of Boltzmans constant k would be 1.0 in units of ergs per new degree.

The Heavy Monatomic Gas Model

The heavy monatomic gases (at temperatures below which changes in electron levels are involved) are similar to the ideal monatomic gas except that the molecules have volume and attract and repel each other. Their only kinetic energy is that of translation and they follow the laws of classical Newtonian mechanics so that quantum mechanics is not needed to interpret their behavior (Lewis and Randall, p. 605, 1961). The internal energy is the sum of the intermolecular potential energy and the translational kinetic energy,

$$E = E_p + E_k \qquad (7)**$$

where E_p is that portion of the internal energy which is potential energy and E_k is that portion which is kinetic energy. The double asterisk indicates equations which apply specifically to the heavy monatomic gas model.

Consider N molecules of such a heavy monatomic gas at temperature T and volume V. Molecule i of this gas will show large fluctuations in its kinetic energy e_{ki} because of exchanges of potential and kinetic energy: (a) within itself, (b) with other molecules, and (c) with the walls of the container. There will be a time average of this kinetic energy of molecule i, $<e_{ki}>$. Consider the numerical average kinetic energy of two molecules $(e_{ki} + e_{kj})/2$. This will still show large fluctuations; however, they will be considerably less; yet, the time average of the kinetic energy per molecule for the two molecules will be the same as the time average of the single molecule.

If the average kinetic energy of all N molecules is taken, the fluctuations of the average kinetic energy will be much less and become zero as N approaches infinity. The average by molecules is then equal to the time average of the kinetic energy of one molecule. Since the kinetic energy of this model of a heavy monatomic gas is equal to the kinetic energy of an ideal monatomic gas,

$$E_k = 3/2\ NkT \qquad (8)**$$

and,

$$T = (2/3k) \sum_{\substack{i=1 \\ N \to \infty}}^{i=N} e_{ki}/N = (2/3k) < \sum_{i=1}^{i=N} e_{ki}/N > = (2/3k) <e_{ki}> \qquad (9)**$$

The intermolecular potential energy (per mole) E_p is a function of an average distance between the molecules, or

$$E_p = f(\bar{r}) \qquad (10)**$$

where $\bar{r}$ is an average of the intermolecular distances weighted for time and the shape of the force distance function.

Internal Pressure and Internal Work

For the model of the heavy monatomic gas the internal energy is the sum of the kinetic energy of the molecules and their intermolecular potential energy by equation (III-7**). Consider that such a gas is placed in a constant temperature bath and then is allowed to expand gradually. The kinetic energy of the molecules will remain constant for the kinetic energy is a function of the temperature only and not the volume (or the weight of the molecules etc.). Refer to the discussion in this chapter on the thermal equilibrium between two different gases through a hypothetical membrane. Hence,

$$(\partial E_k/\partial T)_T = 0 \qquad (11)**$$

Then from (6) and (10),

$$(\partial E_p/\partial V)_T = (\partial E/\partial V)_T \qquad (12)**$$

$(\partial E/\partial V)_T$ is known as the internal pressure (or Pi) because of its analogy to the actual pressure. The integral $\int Pi\ dV$ is known as the internal work for it is the work done against the intermolecular forces when expanding at constant temperature. The total work done in expanding at constant temperature is,

$$\int (Pi + P) \, dV \qquad (13)**$$

The internal pressure has the same dimensions as pressure. When Pi is multiplied by the molar volume V and divided by the number of molecules in one mole N_o it becomes equal to the time average intermolecular force. Hence,

$$Pi \, V = N_o <f> \qquad (14)**$$

where $<f>$ is the time average intermolecular force acting in line of path. The line of path restriction is necessary, for this is the component of the forces acting on the molecules which does work changing the molecules kinetic energy. The component of the forces which is perpendicular to the molecules path does no work. It only changes the direction of the path.

Averages and Thermodynamic Variables

It has been pointed out that both pressure and temperature as measured in the laboratory and hence used in thermodynamics are averages. They can be considered either as time averages over very small units of measure, or as averages taken over a very large number of similar units at the same time. All the thermodynamic properties of real substances can be regarded as either of these types of averages.

The variation of a property about its average is usually not measurable. However, there is one type of system in which that variation becomes very evident — gases near their critical point. These gases have such a large variation in their density from point to point that they are milky in appearance.

It is not necessary from an abstract thermodynamic viewpoint to interpret thermodynamics in terms of averages of molecular properties; however, it does give one a deeper understanding and appreciation of thermodynamics. The proper use of averages requires that one know the different kinds of averages and the reasons for the use of each.

Some Comments on Averages

The three most common types of averages are: the arithmetic mean, the root mean square, and the geometric mean. The correct mean for use in any particular case depends on the mathematical relationships on the property involved.

Consider this use of the arithmetic mean. The total kinetic energy E_k of this model of a real monatomic gas is,

$$E_k = \tfrac{1}{2} \sum_{i=1}^{N} m v_i^2 = \sum_{i=1}^{N} e_{ki} = N e_k \qquad (15)**$$

where the summations over all N molecules must be taken at one instant in time, for the kinetic energy of the individual molecules varies rapidly. The quantity e_k is the arithmetic mean of the kinetic energy of all molecules. This is useful in that one obtains the kinetic energy of N molecules by merely multiplying this arithmetic average of the kinetic energy per molecule by the number of molecules as in (15).

On the other hand, if one wishes to consider the kinetic energy of the molecules in terms of a mean molecular velocity, one must calculate it from the arithmetic mean of the kinetic energy of the molecules. The value obtained is the root mean square of the velocities which is different than the arithmetic mean of the molecular velocities. The expression for this is,

$$\bar{\bar{v}} = ((\sum_{i=1}^{N} v_i^2)/N)^{1/2} \tag{16}$$

and then,

$$E_k = Nm\,\bar{\bar{v}}^2/2 \tag{17}**$$

where the double bar here indicates the root mean square velocity.

For a different example of an average consider that w_i is the probability that an event i will happen. If the various w_i are all of different value and also independent of each other then the probability W that all N events could happen is the product of all the w_i,

$$W = w_1 w_2 \ldots w_N = (\bar{w}')^N \tag{18}$$

where $\bar{w}'$ is the geometric mean of the probabilities.

Certain quantities discussed in the first part of this chapter have been described as time averages. Such averages are arithmetic means so calculated that they are equivalent to averages of numbers taken with equal time intervals between the numbers. If such numbers are taken at different time intervals they must be weighted in proportion to the time intervals between them for the calculation of the time average. The time average is generally the best average for the prediction of the future if there are no time trends. As such it is also called the expected value and indicated by $< >$. Time averages will be utilized in the next chapter.

More Complex Molecular Models

Molecular concepts of temperature, pressure, internal energy, and entropy have been developed for a monatomic gas which follows the laws of classical mechanics. These same concepts can be extended to more complex models of molecules by adding other molecular motions such as rotation, vibration, etc. to each of the molecules. Each of these motions provides a component to the internal energy, and each of the motions has both a kinetic and potential energy component.

These same types of motions also exist in liquids and in solids, however, their area of motion is restricted by intermolecular forces. At the moderate temperatures and pressures encountered in rheology it is convenient to assume that the molecules follow the laws of classical mechanics. We know that this is not strictly true; however it provides a qualitative model for the visualization of the molecular processes which occur in rheology.

The use of models where the molecules follow the laws of classical mechanics is not appropriate for either very high or very low temperatures. At high temperatures energy changes occur with the atoms. At low temperatures the molecular models must be modified to account for the approach to the specific heat of all materials to zero at zero degrees Kelvin and to account for the unusual properties of materials at very low temperatures. Models for both high and low temperatures involve the quantization of energy, and other means for the storage of energy.

Chapter IV

The Two Definitions of Entropy and the Second Law of Thermodynamics

Are the Two Entropies Equivalent?
The statistical definition of entropy is,

$$\Delta S = k \ln W \tag{1}$$

where W is the probability that a body will go spontaneously from state (1) to state (2). "Spontaneously" implies that there is no outside action on that body, i.e., that the body is isolated from its surroundings and hence that the total energy of the body must remain constant. By definition a body can only go spontaneously (on the average) from a less probable state to a more probable state. Hence, the statistical entropy must follow the second law of thermodynamics — the entropy of an isolated system can only increase or remain the same.

The definition of the thermal entropy is

$$\Delta S = \int \delta q / T \tag{2}$$

where q is the heat absorbed. It is evident from this definition that when heat flows from a higher temperature to a lower temperature entropy must increase. Hence, when considering only the flow of heat, the statement of the second law that entropy can only remain the same or increase, and the observation that heat always flows from a higher to a lower temperature are related statements which are consistent with each other.

Does the fact that the thermal and the statistical entropy both obey the second law prove that they are identical quantities? No! For, one entropy could increase while the other remained the same. This problem of the identity of the two definitions of entropy bothered me. I could find no solution in the literature which satisfied me so I decided to try and check one entropy against the other by calculation using the simplest possible examples.

The Status of the Statistical Mechanics of Classical Systems
The calculation of the statistical entropy change of an ideal gas in going spontaneously from state (1) to state (2) is simple. The probability w that one molecule will be in V_2 a part of V_1 is, V_2/V_1. Since the probability that one particular molecule will be in V_2 is independent of the positions of the other molecules, the probability that any molecule is in V_2 is independent of

the positions of the other molecules. Hence, the probability W that all N molecules will be in V_2 at any given instant is the product of all the individual probabilities,

$$W = w^N = (V_2/V_1)^N \qquad (3)*$$

Hence, the entropy change in going from state (1) to state (2) for an ideal gas from equations (1) and (3) is,

$$\Delta S = Nk \ln V_2/V_1 \qquad (4)*$$

Such a spontaneous contraction of an ideal gas would be at constant kinetic energy (i.e., at constant temperature), for there are no intermolecular forces, and hence there can be no interchange of energy between the intermolecular potential energy and the kinetic energy of the molecules.

Real gases are much different since their molecules attract and repel each other, and as a result any spontaneous contraction of a gas from a V_1 to a V_2 would be accompanied by a change in temperature. Since the molecules attract and repel each other the probability that any one molecule is in a V_2 is not independent of the positions of the other molecules. Hence, the probability W is no longer equal to the Nth power of the ratio of the volumes.

I found for the purposes of the proposed calculations, the literature on the statistical mechanics had advanced little beyond the work of Boltzman (1964), who first promoted the relation between probability and entropy.

Most of the structure of the mathematics of classical statistical mechanics is built on an assumption which has not been proved (Tolman p. 65, 1938; Rushbrooke, p. 14, 1949; Eyring, p. 83, 1964 and many others). The validity of the assumption is stated to be shown by the agreement of the theoretical results deduced from that assumption with experimental data. The assumption is stated in various ways by different authors. One statement of this assumption is that phase averages are equal to time averages. Several authors state that the reason that time averages have not been used is that methods were not available to calculate them (Kinchin, p. 52, 1949; Isihara p. 24, 1971; Munster, pp 20, 37, 1969). It must be noted, however, that Kirkwood (1946), Chandrasekhar (1943) and others have used time averages in their statistical mechanical methods for the calculation of steady state transport processes.

The approach taken in this chapter is that the *probability should equal the time average*. The significance and meaning of this statement will be described in detail, and calculations will be made to demonstrate that time averages in the area of statistical mechanics can be calculated in simple cases.

Time Averages and Their Relationship to Probability

The time average is a simple concept — one which is used intuitively without a knowledge of its formal name. Consider a series of readings taken at equal time intervals. Their average is a time average. On the other hand, if a series of readings is taken twice as often in the first hour as in the second hour, the average of all these reading is not a time average; it would be biased, because more weight is given to the readings taken in the first hour. However, a time average can be calculated by giving each reading a weight proportional to the time interval between the readings.

Consider a second example: There are two pockets 'A' and 'B'. A ball is automatically switched back and forth between the pockets according to a fixed schedule, spending ten seconds in pocket 'A' and five seconds in Pocket 'B', then returning to pocket 'A' and repeating the cycle, the ball is in pocket 'A' two thirds of the time; so the time average that the ball is in pocket 'A' is two thirds, and the time average that the ball is in 'B' is one third. If one were to assign the value of one to the ball being in 'A' and zero to the ball being in 'B' and make a plot of the values against time the ordinate would have an average value of two thirds — a time average. If a person thought he could tell when the ball was in pocket 'A' when he was in another room this could be checked by trial. If his prediction was correct approximately two thirds of the time, this does not confirm his claimed ability, for purely random guesses should give the same result. *The probability that the ball is in pocket 'A' at any given random instant is equal to the time average of it being in 'A'.*

Consider a third example: a point oscillating on a line in simple harmonic motion. Select a small segment of that line Δx. The amount of time that the point spends in that segment during one half cycle is the length of that segment divided by the average velocity of that point in that segment or $\Delta x / \dot{x}$. The proportion of time spend in Δx is equal to the probability (w) that it is in Δx at any given random instant is,

$$w = \Delta x / (\dot{x} \tau / 2) \qquad (5)$$

where τ is the time for one cycle.

The sum of all probabilities must equal one, so, equation (5) may be checked by converting Δx to dx and integrating. We find,

$$\int dw = \int dx / (\pi (1-x^2)^{1/2}) = 1 \qquad (6)$$

Expression (5) gives the probability that the point would be within a given segment of its path Δx. The probability that the point would be within a given velocity range $(\dot{x}_2 - \dot{x}_1) = \Delta \dot{x}$ is obtained in a similar way. The amount of time that the point is within the velocity range $\Delta \dot{x}$ is the change of velocity divided by the rate of change of velocity, or $\Delta \dot{x} / \ddot{x}$ where $\ddot{x}$ is the average acceleration within $\Delta \dot{x}$. Hence, the probability that the point would be within the velocity range $\Delta \dot{x}$ at any given random instant is,

$$w = \Delta \dot{x} / (\ddot{x} \tau / 2) \qquad (7)$$

The Probability of Multiple Simultaneous Events

Consider that there are N points traveling independently on separate paths. (No two can have the same motion.) There is a Δx on each of these paths and the probability that each will separately be in its Δx is w. Consider for convenience that the length of the "Δx"'s are such that the "w"'s for all are equal. Since the probabiality that any one point is within its Δx is independent of the position of the other points, the probability W that all N points will simultaneously be within their Δx is,

$$W = w^N \tag{8}$$

Now consider the probability that a point i will be in its Δx if its probability depends on where the other points are on their path. Let w_{ci} be the conditional probability that point i is in its Δx, and ^iW be the probability that all points other than i are in their Δx. Then the probability W that all points are in their particular Δx is,

$$W = {}^iW w_{ci} \tag{9}$$

If all points are equivalent to the extent that both conditional probabilities are equal for all points, then,

$$W = {}^1W w_{c1} = {}^2W w_{c2} = \ldots = {}^NW w_{cN} = (w_c)^N \tag{10}$$

Equations (8) and (10) are very similar, however, they are basically different. For a discussion of conditional probability refer to Feller, Chapter V, Volume I, 1957, though this particular case is not discussed by him.

A Molecular Model Based on Classical Newtonian Mechanics

Consider a monatomic gas whose molecules obey the following rules:

A. The energy is a continuous function.

B. Any force acting on a molecule is equal to the molecular mass times its acceleration.

C. The molecules have a kinetic energy of translation but no rotational or vibrational kinetic energy.

D. The molecules attract and repel each other as a function of the distance between them. The exact form of this function is immaterial. It could be the inverse square law with repulsion on contact, or it could follow Londons 6-12 rule, or any other relationship.

E. The individual molecules have a volume. However, this assumption of volume is not a condition for the following developments.

The molecules of this gas follow the laws of classical mechanics.

When such a gas is isolated in a volume V_1 the molecules move in a cyclic path such that all molecules return to an initial starting point (at a t_o) with a period of τ which approaches infinity. All molecules are identical and follow the same path — they are simply on different parts of the same very long path. Consider that V_2 is a portion of V_1 and that sections of the path which cut through V_2 are "Δx"'s. During the period τ there will be times when all molecules are in V_2. These times will be of very short duration — much less than the time it takes for the fastest molecule to cross V_2. When all molecules but molecule i are in V_2 the proportion of time that molecule i is in V_2 is proportional to the conditional probability w_c that molecule i is in V_2 given that all other molecules are in V_2. Since all molecules are identical the probability W that all molecules are in V_2 is the same as equation (10), or

$$W = (w_c)^N \tag{11}$$

Further relationships can be developed using the probability for molecular velocities as developed in equation (7). Consider again a molecule i on its long cycle τ. The probability w that the molecule will be in a given velocity range is equal to that velocity range divided by its average acceleration in that range, all times the number of times that velocity range is reached (n) per cycle divided by the cycle time τ. From the property B above then,

$$w = \Delta \dot{x} n / \tau \ddot{x} = k/f \tag{12}$$

where k is a proportionality constant and f is an average force in line of path acting on that molecule (within that $\Delta \dot{x}$).

Consider again V_2 a portion of V_1; however, in this case V_2 does not have a particular place or shape. It is any and all of the numerous volumes in V_1 which have a specific size in units of volume. If all molecules are spontaneously in V_2 the molecules will have a different average velocity and a different average intermolecular force than if they were just in V_1. This property comes from item D above, for the intermolecular forces change with distance and the work done with this change in distance will change the average kinetic energy of the molecules.

The same reasoning applies here as in the development of equation (10). Consider that all molecules but i have the average velocity of molecules in volume v_2, then the condition probability w_c that molecule i will also have that average velocity in proportional to $1/\bar{f}$, where $\bar{f}$ *is an average intermolecular force in line of path when all molecules are in* V_2. Since all molecules are alike, the probability that the average velocity of all molecules is that of V_2 is,

$$W = (w_c)^N = k(1/\bar{f})^N \tag{13}$$

where $\bar{f}'$ is an average intermolecular force in line of path, where the type of average is that appropriate for the purpose.†

The probability that all molecules are in V_1 is unity, so,

$$1 = k\,(1/\bar{f_1'})^N \tag{14}$$

dividing equation (13) by (14) we obtain

$$W = (\bar{f_1'}/\bar{f_2'})^N \tag{15}$$

where $\bar{f_1'}$ and $\bar{f_2'}$ are the average intermolecular forces respectively in volumes V_1 and V_2.

The w in equation (12) and the W in equations (13) and (15) are equal to time averages. The exact mathematical expression for the calculation of

---

†See note in Chapter XII, p. 136.

these time averages in terms of molecular velocities and intermolecular forces is not available. However, although different types of averages for a single distribution are not equal to each other, the ratio of one type of average for two similar distributions is approximately equal to the ratio of a second type of average for those same two distributions. The distribution about their means of the intermolecular forces and molecular velocities must remain similar with moderate changes in volume. Hence, we can make the approximation,

$$\bar{f}'_1/\bar{f}'_2 = f'_1/f'_2 \qquad (16)$$

and

$$W = (\bar{f}'_1/\bar{f}'_2)^N \qquad (17)$$

where $\bar{f}'_1$ and $\bar{f}'_2$ are the arithmetic means of the intermolecular forces in line of path.

The Internal Pressure and the Intermolecular Forces

The internal energy of our model of a monatomic gas has only two components, kinetic energy from the translation motion of the molecules and potential energy from the intermolecular attraction. This is expressed in equation III-7

$$E = K_k + E_p \qquad (18)**$$

When a gas is allowed to expand at constant temperature the kinetic energy remains constant and the internal energy is increased only by the work done against the intermolecular forces. This increase in the internal potential energy is called the internal work and is equal to,

$$\int (\partial E/\partial V)_T dV = \int Pi \, dV \qquad (19)$$

where $(\partial E/\partial V)_T$ (or Pi) is called the internal pressure in analogy to the (external) pressure. In such a constant temperature expansion, heat must be absorbed from a constant temperature reservoir and changed into potential energy. The internal pressure can be regarded as produced by that component of the intermolecular forces which act in line of the molecular paths to change the kinetic energy. It is hence proportional to the arithmetic average of the intermolecular force f in equation (17). Hence,

$$f = Pi V/N_o \qquad (20)**$$

where the factor of the molar volume divided by the number of molecules in a mole is introduced to convert the force in terms of pressure to force per molecule. From (1), (17) and (20)

$$\Delta S'' = N_o k \ln (Pi_2 V_2 / Pi_1 V_1) \qquad (21)**$$

where the double prime on $\Delta S''$ indicates the entropy change per mole calculated from expression (21).

Fluctuations and the Equations of State

When a gas spontaneously contracts from V_1 to V_2 the duration of time that all molecules are in V_2 during any one contraction must be very short — much less than the time it takes one molecule to cross V_2. However, these contractions must occur many times during the very long cycle time τ. It is presumed that the average properties of the gas over the many times all the molecules are in V_2 are the same as though it were held by walls in V_2. If this is so, the equation of state $B = f(E, V)$ is valid, and the average properties of the gas during these contractions can be obtained from the tables of the gas properties. (Except for gas pressure as there are no walls to accept pressure.) The validity of equation (21) can then be checked by calculating $\Delta S''$ using the internal pressures calculated from the thermodynamic tables.

Any such spontaneous contraction would occur at constant internal energy E so the entropy changes calculated from (21) would be over a constant energy path.

The Heavy Monatomic Gases

The molecules of the heavy monatomic gases follow the laws of classical mechanics (Lewis and Randall p. 605, 1961). These are considered to be, argon, xenon, and krypton. There are very good thermodynamic tables for argon and less extensive though good tables for xenon. No tables for krypton have been located. The behavior of argon and xenon in the range covered by these tables is far from that of an ideal gas; so, a test of the fit of equation (21) using these tables is not just a superficial test which could be passed by gases which behave similar to an ideal gas.

Calculation — Argon

The thermodynamic tables for argon prepared by Gosman (1969) are judged to be accurate and to cover the range over which equation (21) could be tested. Gossman also gives a sixteen term equation of state from which an expression for P_iv can be derived,

$$\begin{aligned}
P_i v &= (_M E/\partial V)_T V = (T(\partial P/\partial T)_V - P)V \\
&= -\rho\,(n_2 + 2n_3/T + 3n_4/T^2 + 5n_5/T^4) - \rho^2 n_7 \\
&\quad - \rho^2\,(3n_9/T^2 + 4n_{10}/T^3 + 5n_{11}/T^4)\exp(-n_{16}\rho^2) \\
&\quad - \rho^4\,(3n_{12}/T^2 + 4n_{13}/T^3 + 5n_{14}/T^4)\exp(-n_{16}\rho^2) \\
&\quad - \rho^5 n_{15}
\end{aligned} \tag{22}$$

where ρ is the density of the argon and the values of the constants n_i are given by Gosman.

The average properties of argon during their spontaneous contractions can be obtained from Gossman's tables by interpolating over constant energy paths. The interpolated values previously obtained by Hull (1973) were used for these calculations.

Tables I and II show the values of $\Delta S''$ calculated over two constant energy paths. The pressure range from 0.01 atmospheres to the highest pressure listed in the thermodynamic tables for each path. The values for the entropy change obtained from (21) are compared with Δs the entropy change interpolated from the thermodynamic tables of Gosman. Gosman's values for entropy were obtained by the methods of classical thermodynamics.

The path in Table I passes closest to the critical region where the behavior of argon is furthest from ideal. The calculated values of $\Delta s''$ are generally within two percent of Δs. Table II shows the second path at a higher constant internal energy. The agreement is better presumably because the path is further from the critical region.

Table I.
Comparison of the Entropy Change for Argon Gas as Calculated from Equation (21) to the Entropy Change Interpolated from Thermodynamic Tables. Constant Internal Energy of $E = 9034.6$ J/mol.

Pressure atm.	Temperature °K	Density mol/liter	Piv cc-atm/mol	$\Delta s''$ J/mol°K	Δs J/mol°K	Difference %
0.01	107.78	0.0011389	2.7912	−38.20	−38.12	+0.21
0.04	107.85	0.0045234	11.161	−26.67	−26.60	+0.26
0.10	107.98	0.011306	27.893	−19.06	−18.99	+0.36
0.50	108.88	0.056459	138.96	−5.71	−5.69	+0.35
1.00	110.00	0.11272	276.56	0	0	
2.00	112.19	0.22467	548.84	5.70	5.60	+1.78
5.00	118.42	0.55542	1317.2	12.98	12.74	+1.89
10.00	127.74	1.0891	2463.4	18.18	17.79	+2.19
20.00	143.26	2.0920	4235.0	22.86	22.38	+2.14
50.00	175.58	4.6435	7485.2	27.95	27.53	+1.53
100.00	208.26	7.7589	11360.	31.02	30.71	+1.03
200.00	245.64	11.702	15840.	33.65	33.41	+0.71
300.00	268.88	14.227	19188.	35.10	34.87	+0.65
500.00	299.07	17.531	22866.	36.71	36.68	+0.08

Table II.
A Second Comparison of the Entropy Change for Argon Gas as Calculated from Equation (21) to the Entropy Change Interpolated from Thermodynamic Tables. Constant Internal Energy of $E = 9544.4$ J/mol.

Pressure atm.	Temperature °K	Density mol/liter	Piv cc-atm/mol	$\Delta s''$ J/mol°K	Δs J/mol°K	Difference %
0.01	148.66	0.0008198	1.6905	−38.21	−38.21	0
0.04	148.69	0.0032796	6.7622	−26.69	−26.69	0
0.10	148.78	0.0081975	16.902	−19.07	−19.09	−0.10
0.50	149.30	0.040955	84.444	−5.70	−5.72	−0.34
1.00	150.00	0.081826	167.56	0	0	
2.00	151.34	0.16329	332.53	5.70	5.69	+0.19
5.00	155.24	0.40558	810.98	13.11	13.09	+0.19
10.00	161.37	0.80225	1557.7	18.54	18.50	+0.18
20.00	172.40	1.5665	2887.7	23.67	23.64	+0.13
50.00	198.59	3.6307	5955.2	29.69	29.75	−0.20
100.00	228.78	6.3865	9391.8	33.47	33.67	−0.59
200.00	266.32	10.195	13786.	36.67	37.00	−0.89
300.00	290.50	12.761	16808.	38.31	38.75	−1.12

Calculations — Xenon

Tables of the thermodynamic properties of xenon have been published by Michels (1956). An equation of state is not given and the temperature and pressure ranges are less than in the tables for argon.

Calculations were made for $\Delta s''$ for one constant energy path and the results are shown in Table III. Michels' tables were interpolated by fitting a second degree polynomial to the four closest points, using the method of least squares. The polynomial was tested for fit, and if not satisfactory transformations were tried until a satisfactory fit was obtained. The partial derivatives needed for the calculations of the internal pressure (by means of equation 19) were obtained from the slope of the polynomials used for the interpolations.

The difference between the calculated values of $\Delta s''$ from equation (21) and Δs interpolated from Michels thermodynamic tables is approximately the same as obtained for argon in Table II.

The fit of equation (21) was also tested for neon and nitrogen over similar constant energy paths. The difference between $\Delta s''$ and Δs ranged to 7% for neon and to 10% for nitrogen. The poor fit for neon is attributed to its not being a heavy monatomic gas whose molecules follow the laws of classical mechanics. The poor fit for nitrogen is attributed to its being a diatomic gas.

Table III.
Comparison of the Entropy Change for Xenon Gas as Calculated from Equation (21) to the Entropy Change Interpolated from the Thermodynamic Tables of Michels. Constant Internal Energy of $E = 0$.

Density Amagats	Temperature °K	Piv cc-atm/mol	$\Delta v''$ J/mol°K	Δs J/mol°K	Difference %
1.00	273.15	299.04	—	—	
14.41	303.15	3947.1	21.44	21.48	−0.18
38.29	348.15	9199.9	28.48	28.56	−0.35
85.42	423.15	17701.9	33.92	34.04	−0.34

Discussion and Summary

A model of a heavy monatomic gas whose molecules follow the laws of classical mechanics was described. The model was then used to develop an expression for changes in entropy over constant energy paths using the definition of statistical entropy, $\Delta S = k \ln W$. The probability W is assumed to be equal to a time average.

A reasonable agreement was obtained between the calculated values of entropy change (from (21)) and the entropy change interpolated from the thermodynamic tables for two gases — argon and xenon. Perfect agreement would not be expected as there are undoubtedly small errors in the original measurements from which all quantities were calculated, and there is one approximation used in the development of (21), i.e., that of the proportionality of the averages in expression (16).

The fact that the data for neon and nitrogen do not fit equation (21), while that for argon and xenon do agree with the concepts of the model. A further indication of the validity of the model is the good fit obtained even though the behavior of the gases is far from ideal. The possibility that an error exists in the development of the equation and that a fit is obtained by chance is quite remote.

Equation (21) is a partial solution to the classical many body problem in which an expression is sought to describe the behavior of a very large number of bodies which follow the laws of classical mechanics.

The equality of the time average and probability has been frequently stated in the literature; however, to my knowledge no adequate proposal for obtaining the time average has previously been proposed for models based on Newtonian mechanics. However, in the area of quantum mechanics the methods of weighting effectively give time averages. There is a close similarity between the weighting methods of quantum mechanics and the weighting methods used here for Newtonian systems.

It should be possible to develop methods for combining the equations developed here for the calculation of the entropy of Newtonian systems which use the equations of quantum mechanics. However, a major problem which must be solved before this is done is to develop a method for the partitioning the effects of the various molecular movements on the internal pressure. For example, with a diatomic molecule the rotational and the vibrational movements of the molecules could contribute to the internal pressure.

It should be noted that the two volumes in equation (21)** for the entropy of a monatomic gas are inverted with respect to the two volumes in equation (4)* for the entropy of an ideal gas. Equation (21)** should not be considered a modification of equation (4)*. When the interatomic forces approach zero (the ideal gas condition) equation (21)** becomes indeterminate.

Chapter V
The Thermodynamic Functions

The properties of state, pressure, volume, temperature, entropy, and internal energy are best described as primary properties of state for they are inherent. The thermodynamic variables known as functions of state (Gibbs, G; Hemholtz, A; Enthalpy, H; etc.) are best described as secondary variables of state, for they are defined in terms of the primary properties. It is best not to think of these functions as properties in themselves but only as intermediate quantities, useful for certain further calculations. Their purpose is to shorten and simplify calculations. In other words they are a very useful mathematical "trick". Note that it is possible to describe all thermodynamic relationships without the use of the thermodynamic functions by replacing them with their algebraic equivalent. Bridgman has developed this concept further by developing a table for thermodynamic relationships (for two independent variable thermodynamics) and expressing them in terms of P, T, S, and the three partial derivatives which are most readily measured — $(\partial V/\partial T)_P$, $(\partial V/\partial P)_T$, and $Cp = (\partial H/\partial T)_P$ (Lewis and Randall, Appendix 6, 1961).

Certain names which are sometimes used for these functions imply that they are a measure of a type of energy (i.e., heat content for enthalpy, and free energy for the free energy functions). Specific differences between specified values of these functions are the *desired* meaningful energies. These differences are designated by a "Δ". It also happens tht there are some very useful partial differential relationships between the functions and other thermodynamic properties.

The Enthalpy and Analogous Functions

The enthalpy is defined by,

$$H = E + PV \tag{1}$$

It is a secondary variable of state for it is defined in terms of E, P, and V primary variables of state. It is a function created as a convenience for making calculations when the pressure of a system is held constant. Differences in enthalpy at the same pressure are the amounts of heat added or removed in going from one thermodynamic state to another. Differences in enthalpy at different pressures do not have the same simple meaning.

Differences in enthalpy at constant pressure are labeled ΔH. The heats of melting, the heats of evaporation, and the heats of chemical reactions are labeled ΔH but are also at constant temperature. ΔH is also used to designate the heat absorbed in heating at constant pressure but at varying temperatures. In order to emphasize which variables are held constant a superscript will be added. For example, ΔH^P designates that P is held constant, while ΔH^{PT} designates that both P and T are held constant. An example of the application of this notation is the Clapeyron equation for the change in vapor pressure of a substance with change in temperature. It is then written,

$$dP/dT = \Delta H^{PT}/T\Delta V^{PT} \qquad (2)$$

where ΔH^{PT} is the heat of vaporization or evaporation and ΔV^{PT} is the corresponding change in volume. The superscript "PT" emphasizes that the right portion of the equation refers to constant pressure and temperature conditions although the left hand refers to the rate of change of pressure with respect to increasing temperature. This initiates a notation which may seem over elaborate at this point but which will become more useful in later chapters when more complex relationships are involved.

One of the simplest uses of the enthalpy is that its partial derivative with respect to temperature (pressure constant) is the specific heat at constant pressure,

$$Cp = (\partial H/\partial T)_P = (\partial E/\partial T)_P + (\partial V/\partial T)_P \qquad (3)$$

The manipulation of differentials obtained from the definition of enthalpy can provide some useful relationships.

The total differential of H from its definition is,

$$dH = dE + P\,dV + dP \qquad (4)$$

From the first law,

$$0 = dE + P\,dV - T\,dS \qquad (5)$$

Taking the difference between (4) and (5),

$$dH = V\,dP + T\,dS \qquad (6)$$

By keeping S constant in (6) we obtain,

$$(\partial H/\partial P)_S = V \qquad (7)$$

By keeping P constant in (6) we obtain,

$$(\partial H/\partial S)_P = T \qquad (8)$$

The enthalpy as defined above presumes there are only two independent variables of state, P, and T. When more than two independent variables of state are involved, functions analogous to the enthalpy can be defined and are useful. The symbol for these functions will be HH where the extra "H" is symbolic of the additional variables,

$$HH = E + PV + XY + \ldots \qquad (9)$$

Extended thermodynamic functions of the above type are mentioned in Lewis and Randall (p. 514, 1961) but it is stated they have not been found useful except in the case of magnetic fields. The writer has found them invaluable in understanding the thermodynamics of elastic deformation. Although it is easy to extend the definition of HH to any number of independent variables of state, it will be understood that only three are intended unless otherwise stated.

The various thermodynamic relationships for HH can be developed by the same methods used for H. The following are examples analogous to (4) thru (8) above.

The total differential of HH from its definition (9) is,

$$dHH = dE + P\,dV + V\,dP + X\,dY + Y\,dX \qquad (10)$$

From the first law we obtain,

$$0 = dE + P\,dV - T\,dS + X\,dY \qquad (11)$$

Taking the difference between (10 and (11),

$$dHH = V\,dP + T\,dS + Y\,dX \qquad (12)$$

By keeping different variables constant in (12) we obtain,

$$(\partial HH/\partial P)_{SX} = V \qquad (13)$$

$$(\partial HH/\partial S)_{PX} = T \qquad (14)$$

$$(\partial HH/\partial X)_{PS} = Y \qquad (15)$$

As a second example consider a rubber band held at one end and suspending a weight on the other. This is a system at constant pressure P, and constant force in tension X. The length of the rubber band is Y. When this rubber band is heated some of the added energy will go into increasing the internal energy in the amount dE, some into work against the atmosphere when the rubber band increases in volume in the amount $P\,dV$, and some into lifting the weight on the end of the rubber band in the amount of $X\,dY$, for a stretched rubber band suspending a weight will shorten when heated. The sum of these quantities is the specific heat of the rubber band at constant pressure and constant tension. The expression for this is,

$$Cpx = (\partial HH/\partial T)_{PX} = (\partial E/\partial T)_{PX} + P(\partial V/\partial T)_{PX} + X(\partial Y/\partial T)_{PX} \qquad (16)$$

Other relationships can be developed for HH by methods parallel to those used for H.

The Free-Energy Functions

The free-energy functions of primary interest are:

The Hemholtz free-energy function, $A = E - TS$, $\qquad (17)$

The Gibbs free-energy functions, $G = E - TS + PV$, $\qquad (18)$

The GGibbs free-energy function,

$$GG = E - TS + PV + XY \qquad (19)$$

In the latter the extra "G" in both the name and the symbol is symbolic of the extra variables of state.

These free-energy functions are secondary variables of state for they are defined in terms of the primary variables of state P, V, T, S, E, X, and Y. These functions are so defined that differences under specific conditions define the maximum amount of work available from a system in going from a certain state (1) to a specifically related state (2) (i.e., the free energy).

The relations between the two states in the case of the Helmholtz free-energy function is that they both be at the same temperature and volume. For the Gibbs free-energy function it is that they both be at the same temperature and pressure, and for the GGibbs free-energy function it is that they both be at the same pressure, temperature, and X. These conditions are the restraints on the system, and the free energy function which is appropriate to use is that whose free energy restrains match most closely that of the real system to which the calculations are to be applied.

The most useful of the free-energy functions is the Gibbs, for constant temperature and pressure are the commonest experimental or industrial conditions. The GGibbs free-energy function has had very little use, but further uses will be developed in later chapters.

A typical use of the Gibbs free-energy function is the calculation of the maximum work available in going by chemical reaction from a state (1) to a state (2) at the same pressure and temperature. The measure of the available work is obtained by taking the difference between the expression for G in the two states. This gives,

$$\Delta G^{PT} = \Delta E^{PT} - T \Delta S^{PT} + P \Delta V^{PT} \tag{20}$$

where the superscripts again emphasize the constant pressure and temperature conditions. Equation is also an expression of the first law; for, ΔE^{PT} is the change in energy within the body (i.e., change in internal energy), $T \Delta S^{PT}$ is the heat absorbed by the body during the reaction, and $P \Delta V^{PT}$ is the work done against the atmosphere. The total of the three is the negative of the amount of work available from the reaction. Hence $\Delta G = -w$. The negative sign sometimes causes confusion. This is caused by the quantity ΔG referring to the changes within the thermodynamic body. A decrease within the body means a release of energy from the body.

The uses of the other two free-energy functions correspond to the above for the appropriate restraints, i.e., constant V and T for ΔA, and constant P, T, and X for ΔGG.

The superscripts of the Δ-quantities of the free-energy functions are not necessarily always the same. However, if they are not the same as listed above, it is best to look into the reason for the difference and to question as to whether the correct free-energy function is being used.

In conventional thermodynamics the third variable of state is usually ignored. This is equivalent to the assumption that the third intensive variable (X) and the third extensive variable (Y) are zero. If the third extensive variable of state is not zero but is held constant, the thermodynamic equations are nominally the same as for only two independent variables of state; however, note that the third intensive variable of state (X) will not remain constant, and the thermodynamic properties depend on the value at which Y is fixed.

The Algebraic Signs of Tension and Compression

The term XY is analogous to PV in the thermodynamic functions. It is best to adopt a sign convention which is also analogous to that used with pressure and volume. When X represents tension or compression the signs of X and Y are then such that the *work done* by an elastic body on its surroundings with a change in deformation is positive (as it is with PV). Hence, it is convenient to use the convention that a compressional force is positive and a tensional force is negative. Hence, a decrease in the amount of compression and an increase in the amount of extension are both positive changes in Y. With this convention the term XY will have the correct sign for both tension and compression.†

The Fugacity

The Gibbs free-energy function is used for the definition of fugacity. For a constant temperature condition the fugacity is defined by the equation,

$$\Delta G^T = (G_2 - G_1)^T = RT \ln (f_2/f_1) \qquad (21)$$

where G is the molal Gibbs free energy function and f is the fugacity. Note that the pressure is not specified as constant and that only T is constant. For an ideal gas the fugacity is equal to its pressure and for most substances it is roughly equal to their vapor pressure.

It is interesting to note what happens when one attempts to develop a function analogous to the fugacity for the GG free-energy function. The total differential of GG is,

$$dGG = dE - T\,dS - S\,dT + P\,dV + V\,dP + X\,dY + Y\,dX \qquad (22)$$

For an ideal substance (such as an ideal rubber as described in Chapter VII) held at constant temperature and constant pressure the term $-S\,dT$ and $V\,dP$ are zero. The terms dE and $P\,dV$ are zero in an ideal rubber. The terms $-T\,dS$ and $X\,dY$ are of equal magnitude but opposite in sign. This leaves,

$$dGG = Y\,dX \qquad (23)*$$

If Y is some theoretical function of X for such an ideal substance the equation can be integrated and an equation analogous to (21) can be developed. However, there is no theoretical relationship for the ideal rubber (as there is between P and V in an ideal gas) so a function analogous to fugacity has not been developed for the third variables of state.

---

†The quantity X is the retractive force in an elastic deformation. It will frequently be referred to as stress though more precisely it is the negative of the stress.

The Chemical Potential

For a closed system with two independent variables of state we have from the first law,

$$dE = T\,dS - P\,dV \tag{24}$$

For an open system in which components i may be added this becomes,

$$dE = T\,dS - P\,dV + \sum_i \mu_i\,dm_i \tag{25}$$

where μ_i is the chemical potential of component i. It is,

$$\mu_i = (\partial E/\partial m_i)_{SVYm_j} \tag{26}$$

where the Y is added as a constant for this is assumed in (27) and (28) though it is not indicated.

The total differential of G is,

$$dG = dE - T\,dS - S\,dT + P\,dV + V\,dP \tag{27}$$

Combining equations (25) and (27) we obtain,

$$dG = -S\,dT + V\,dP + \sum_i \mu_i\,dm_i \tag{28}$$

By combining the total differentials of A and H with equation (27) we obtain,

$$dA = -S\,dT - P\,dV + \sum_i \mu_i\,dm_i \tag{29}$$

and

$$dH = T\,dS + V\,dP + \sum_i \mu_i\,dm_i \tag{30}$$

For open systems with three independent variables of state equation (27) becomes,

$$dE = T\,dS - P\,dV - X\,dY + \sum_i \mu_i\,dm_i \tag{31}$$

In the same way the total differentials of GG and HH may be combined with equation (31) to give,

$$dGG = -S\,dT + V\,dP + Y\,dX + \sum_i \mu_i\,dm_i \tag{32}$$

$$dHH = T\,dS + V\,dP + Y\,dX + \sum_i \mu_i\,dm_i \tag{33}$$

From equations (25), (28), (29), (30), (32) and (33) we obtain six expressions for the chemical potential,

$$\mu_i = (\partial E/\partial m_i)_{SVYm_j} = (\partial A/\partial m_i)_{TVYm_j} = (\partial H/\partial m_i)_{SPYm_j}$$
$$= (\partial G/\partial m_i)_{TPYm_j} = (\partial GG/\partial m_i)_{TPXm_j} = (\partial HH/\partial m_i)_{SPXm_j} \tag{34}$$

THE THERMODYNAMIC FUNCTIONS

The above is another demonstration of the relative ease which the concepts of two independent variable thermodynamics may be converted to the thermodynamics of three independent variables. It is not known how useful these two new equations for the chemical potential may be in rheology or other related areas. It is worth while, however, to explore these various concepts and then see which proves to be useful. Note for later reference the following relationship between the Gibbs function and the chemical potential,

$$\Delta G^{PT} = \sum_{1}^{n} \mu_i x_i \qquad (35)$$

which for a body of one component becomes,

$$\Delta G^{PT} = \mu \qquad (36)$$

Chapter VI
Models of Systems in Balance

Systems and Bodies

A free body diagram in mechanics is the diagram of an isolated body in which all mechanical forces are represented as vectors and couples acting on a body. If the sum of all the static vectors and static couples acting on that body is zero the body is at equilibrium. If one or both the sums is not zero the body is not at equilibrium and motion occurs.

Concepts of thermodynamic systems are similar to the free body diagrams of mechanics in that they represent a body (or bodies) isolated from its surroundings except for certain specified actions by that environment. If the thermodynamic system is at equilibrium no changes will occur. If the system is not at equilibrium changes are taking place or may potentially take place.

In thermodynamics there are a minimum of three types of agents acting on the thermodynamic body each with its associated action (instead of only one in mechanics):

1. The mechanical forces acting on the system from the outside. In thermodynamics one is not normally concerned with the nonbalance of forces which results in motion of the body as a whole. One is concerned with the nonbalance which causes deformation of the body and *associated temperature changes*. (An example of this is the compression of a gas.) Hence thermodynamics is concerned with the stresses and strains within the thermodynamic body and the associated thermal effects. (The thermal effects associated with strain are neglected in mechanics.)
2. Heat transfer. Thermodynamics is concerned with heat transfer as a transfer of energy and its effects on the thermodynamic body as a result of temperature change. It is not concerned with the rate of heat flow (except in the thermodynamics of the steady state).
3. Internal actions which are wholly within the thermodynamic bodies and have no link with the surroundings. These include chemical reaction, temperature and concentration differences, etc. within the thermodynamic bodies.

Thermodynamic systems may also be extended to include variables other than the above three, such as magnetic and electric fields, gravitational fields, surface effects, etc. The conventional treatment of these other variables sometimes does not include the associated thermal effects, which omission can be the source of significant errors.

The terms "thermodynamic system" and "thermodynamic body" have been used without definition; however, the distinction between them is important. The thermodynamic system includes the thermodynamic body plus its environment. The environment includes such items as constant temperature baths and sources of constant pressure. The *system* obeys the first and second laws of thermodynamics; i.e., the energy of a system is constant and the entropy of a system can only remain the same or increase. The energy or the entropy of a *body* can be changed in any arbitrarily desired way. The body is a part of a system.

The various systems of thermodynamics and mechanics are models or idealizations of real systems. The selection of the system model or idealization which best represents (or approximates) a particular real system, within the accuracy desired, requires an intimate knowledge of the behavior of that real system as well as a knowledge of the various possible idealizations which could represent that real system. The final selection can be a matter of judgement, and it may be worthwhile to compare several system models.

Equilibrium and the States of Balance

A dictionary defines "equilibrium" as a static or dynamic state of balance between opposing forces or actions. In thermodynamics three levels in the state of balance are recognized, but only one is designated as thermodynamic equilibrium. The three levels are:

1. *Thermodynamic equilibrium* in which no change or flow of energy or matter is occurring or could occur. No spontaneous change, or change by means of a catalyst, is possible. The system is completely "dead". According to the second law all spontaneous changes must be towards thermodynamic equilibrium.
2. *The metastable state* in which no flow of matter, or flow of energy, or release of energy is occurring, but in which there is a potential change which can occur spontaneously or with the aid of a catalyst. Examples of systems in the metastable state are: superheated or supercooled liquids, mixtures of hydrogen and oxygen, explosives, and some solids in the frozen glassy state.
3. *The steady state* in which energy and/or matter is flowing through the thermodynamic body, though no changes (with respect to time) are occurring in the state of the body itself. This is the subject of a later chapter.

The equations of state described in Chapter II apply to any real body in any of the above three "states of balance". These equations of state can be visualized as descriptions of models in which the independent variables are varied at will and the dependent variables fall where they may.

These independent variables of state can also be thought of as restraints on the system. There must then be a minimum of two restraints, and for a body in the steady state a restraint can be a gradient imposed on a body.

There is sometimes a choice on which variables can be selected as the independent variables. For example,

$$E = f(V, S, m) \tag{1}$$

and

$$S = f(E, V, m) \tag{2}$$

are both legitimate equations of state.† They both apply to what is essentially the same equilibrium state but they apply to different models of systems. They are also related to two different criteria of equilibrium. See equations (3) and (11) following. The preferred equation of state is determined by which of the two most closely represents the visualized physical model. This is generally equation (2).

One might expect there would be a unique condition for thermodynamic equilibrium (and its analogue the metastable state). However, this is incorrect; for, as equilibrium is defined, the conditions for equilibrium depend upon which thermodynamic system (or model) is under consideration. Since the thermodynamic systems are idealistic representations of real physical systems, the selection of a particular idealistic system to represent any real physical situation is somewhat arbitrary. Hence, any condition for equilibrium in any real situation is correspondingly arbitrary.

Guggenheim (p. 29) lists six alternative conditions for equilibrium. These will be discussed individually.

Equilibrium in an Isolated System

One of Guggenheim's conditions for equilibrium is,

for any given E^Σ and V^Σ, that S^Σ is a *maximum* (3)

where the thermodynamic body in the system may contain more than one phase and the Σ superscript indicates the sum of the "E"s, "V"s and "S"s of the phases.

When expression (3) is applied to a single phase at equilibrium it implies the equation of state,

$$S = f(E, V, m) \tag{4}$$

Equation (4) describes a system in which a body is held at constant E and V. This implies a model of a body held in a perfectly rigid container surrounded by perfect insulation.

Such a system is an idealization for materials are not available to construct such a system. It is however, a convenient and useful concept, for it is an approximation for some real systems in which the volume changes are small and heat exchanges with the environment are held low.

---

†The "m" is added to show that the internal energy and the entropy also depend on the amount of material in and the composition of the body.

One can also visualize the whole universe as an isolated system. Clausius did this in making the statement, "Die Energie der Welt ist konstant; die Entropie der Welt strebt einem Maximum zu".† This fueled the debate in his time on running down of the universe with the universe reaching equilibrium at maximum entropy. At equilibrium there would be no source of energy to sustain life and the universe would be dead. Of course the development of the knowledge of atomic and nuclear energy has altered the concept of an imminent "dead universe".

Equilbrium at Constant T and V

Guggenheim lists a second alternate condition for equilibrium,

for any given T and V^Σ, that A^Σ is a *minimum* (5)

where A^Σ is the sum of the Hemholtz free energy functions for all the phases present.

When expression (5) is applied to a single phase at equilibrium it implies the equation of state,

$$A = f(T, V, m) \qquad (6)$$

Equation (6) applies to a body at equilibrium held at constant T and V. The model is a body held in a perfectly rigid heat conducting container, immersed in a constant temperature bath. This system is also an idealization for constant volume containers are not available and "constant" temperature baths only operate between finite limits.

One can conceivably convert the system of constant T and V to an isolated system by placing the bath (containing the body) within a perfectly rigid and perfectly insulated container. The bath would then have no heating or cooling input or controlling device. For it to function as a constant temperature bath it would have to be infinitely large as compared to the body immersed in that bath. However, such a conversion to an isolated system causes the condition of equilibrium to change to that of an isolated system — i.e., maximum entropy of the whole system (the body plus the bath).

Equilibrium at Constant P and T

Guggenheim lists a third alternate condition for equilibrium,

for any given P^α's and T, that G^Σ is a *minimum*
where P^α is the pressure of phase α. (7)

When expression (7) is applied to a single phase it implies the equation of state,

$$G = f(P, T, m) \qquad (8)$$

and the model of a thermodynamic body held at constant pressure and constant temperature. This model comes closest to the actual conditions which

---

†The energy of the world is constant; the entropy of the world rises toward a maximum.

exist in the laboratory or industrial plant, where constant pressure is exerted by the atmosphere and constant temperature is approximated by that of the environment if not by a constant temperature bath.

Since a minimum G^{PT} is the requirement for equilibrium, any spontaneous change at constant P and T must be accompanied by a decrease in G^{PT}, i.e., a negative ΔG^{PT}. Hence, the sign of ΔG^{PT} is the test of whether a chemical reaction or a change in state can occur. For a discussion of free energy under conditions which are not constant P and T see Chapter VII.

The assumption of constant P and T is common in thermodynamics. For example, it is assumed in the calculation of the equilibrium constant K from the expression,

$$\Delta G^{(o)PT} = - RT \ln K \qquad (9)$$

where $\Delta G^{(o)PT}$ is the change in the Gibbs free energy function with each substance in its standard state at pressure P and temperature T. One should be aware of this assumption and although exceptions from this assumption are rare in conventional thermodynamics the assumption may change in three independent variable and steady state thermodynamics.

Adiabatic Conditions at Constant Pressure

Guggenheim lists as a fourth condition for equilibrium,

for any given S^Σ and P^α's that H^Σ is a minimum $\qquad (10)$

This implies the equation of state,

$$H = f(S,P,m) \qquad (11)$$

This is the condition for equilibrium when no heat is transferred to the surroundings and the pressure is constant, i.e., adiabatic conditions at constant pressure. This is an important condition for it is the criteria for equilibrium when equilibrium is approached so rapidly that there is not sufficient time for heat to be transferred to the surroundings.

Equilibrium Under Other Conditions

Guggenheim lists two other alternative criteria for equilibrium,

for any given S^Σ and V^Σ, that E^Σ is a minimum $\qquad (12)$

for any given H^Σ and P^α's, that S^Σ is a maximum $\qquad (13)$

Although these criterion are certainly algebraically correct, I have difficulty in constructing mental models in which the two given variables are held constant and the third variable varied to find the condition in which it is a maximum or a minimum. For this reason I reject (12) and (13) as not being useful for visualizing models at equilibrium or in the metastable state.

Equilibrium with Three Independent Variables of State

The discussion of equilibrium with three independent variables of state will first consider bodies of a single phase only. The same notation will be used as in the previous chapters using X as the third intensive variable of state and Y as its conjugate extensive variable. It is also convenient to use rubber as an example of a thermodynamic body with a tension of X and with its conjugate extension as Y.

For each of Guggenheim's six conditions for equilibrium either X or Y can be added as the third variable of state so there are twelve corresponding conditions for equilibrium. The two most useful are,

$$\text{for any given } P, T, \text{ and } Y, \text{ that } G \text{ is a minimum} \qquad (16)$$

$$\text{for any given } P, T, \text{ and } X, \text{ that } GG \text{ is a minimum} \qquad (17)$$

with their corresponding equations of state,

$$G = G(P,T,Y,m) \qquad (18)$$

$$GG = GG(P,T,X,m) \qquad (19)$$

The model of (16) and (18) is a rubber body held at constant extension in a constant temperature constant pressure bath.

The model of (17) and (19) is similar except that the rubber body is held at constant tension, for example by having it suspend a weight.

These two conditions correspond to (7) in Guggenheim's list. It is an interesting exercise in the manipulating of thermodynamic concepts to develop the other ten corresponding conditions for equilibrium and try to develop mental models for each.

When two phases are present one phase may be subjected to a stress that does not exist in the other. One example would be, a stretched rubber immersed in a solvent. Another example is, a stretched cellulose fiber in equilibrium with water vapor. It would seem that the required work function for equilibrium would be the sum of different ones for each phase. This will be discussed further in a section on swelling in Chapter VII.

Equilibrium in Micro Systems

The properties of a thermodynamic body at equilibrium are normally considered to be unchanging; however, they are constantly fluctuating. With large bodies the variations are so small they are normally ignored. With very small bodies the fluctuations are much larger; however, the *time averages* of the properties of both the large and the small bodies are equal, and these are properties of state of conventional thermodynamics. The properties of the large bodies are referred to as macro and the properties of the very small bodies are referred to as micro.

The best example of a micro property is the pressure (on a very small area) created by the bombardment of the walls of a container by gas molecules. When the walls of the container are heat conducting the internal energy E and even the temperature T are time average properties. (See Chapter III)

Reversible Paths

Reversible paths are an idealized series of continuous changes which never can be attained but only approximated by real systems. Each point on the paths is at an equilibrium or metastable state (by the Guggenheim's definitions which use two independent variables of state). The criterion for a path being reversible is that the total entropy of the *whole system* shall not change over the path. The entropy may flow from one part of the system to the other but the sum of the entropies of all the components of the system must remain constant.

As an example, consider the Carnot cycle of the ideal heat engine. One cycle consists of four reversible paths in series. During the isothermal expansion an amount of heat Q_1 is absorbed from a high temperature source at temperature T_1 by the medium. During the isothermal compression a portion of this heat Q_2 is discharged to the low temperature sink at temperature T_2. The difference $(Q_1 - Q_2)$ is the amount of thermal energy converted to work. The relationships are such that the amount of entropy transferred from the high temperature source (Q_1/T_1) is equal to the amount of entropy transferred to the low temperature sink (Q_2/T_2), that is,

$$Q_1/T_1 = -Q_2/T_2 \qquad (20)$$

There are no entropy changes on the two adiabatic paths so the entropy changes sum to zero, and the entropy of the system is constant.

In most descriptions of the Carnot Cycle the body used for the reversible cycles is a gas; however, equivalent cycles can be obtained with other materials. A Carnot equivalent cycle using rubber for the reversible cycles in place of the gas is described in Chapter VIII.

Real Paths

All real paths of mechanical and thermodynamic systems contain an irreversible component, and the measure of the amount of irreversibility is the increase in entropy of the *system*. That is, the sum of all the entropy changes along any real path must be greater than zero.

For example, any flow of heat requires that a temperature difference be present. No matter how close the two temperatures, there is an increase in entropy equal to $Q/T_2 - Q/T_1$ where Q is the amount of heat transferred.

Also, any movement in the system which involves friction increases entropy by the amount of Q/T where Q is the amount of heat generated by the friction. The same relationship is valid when the flow of electricity occurs and Q is the heat generated by the flow of electricity through a resistance.

Adiabatic Paths

An adiabatic path is one by which energy is being transferred to or from a body by any reversible process except heat transfer. Hence, by the thermal definition of entropy, bodies on adiabatic paths are at constant entropy. If friction or other irreversible processes are involved the path is not adiabatic.

Adiabatic systems can be visualized as surrounded by thermal barriers; however, from a practical viewpoint most processes that approximate adiabatic conditions do so because they take place rapidly, and time is not sufficient for heat to be transferred to or from the surrounding. Examples of processes which may approximate the adiabatic are, the rapid expansion or compression of a gas in a piston engine, and the compression and decompression during the passage of a sound wave.

When there are only two independent variables of state an adiabatic path is an expansion or compression in which the reversible work is equal to the integral of $P\,dV$. When there are three independent variables of state, changes of volume usually occur so that the reversible work is the sum of the integrals of $P\,dV$ and $X\,dY$. Examples of three variable adiabatic processes are, the rapid stretching of rubber, and adiabatic magnetization (as used in magnetic cooling at very low temperatures).

States of Non-Balance

There are two conditions or requirements for balance in a body or in a thermodynamic system:

(a) There must be no spontaneous changes occurring (on a macro basis). An equivalent statement is that entropy of the *body* in the system must not be increasing.

(b) The state must be repeatable no matter from which direction the particular state is approached. An equivalent statement is that the equation of state $B = f(P,T,X...)$ must be valid.

Care must be taken to distinguish between the body and the system in applying the term "state of balance". A body in a steady state system may be in balance while the system as a whole is not. Consider for example a body at the steady state in a temperature gradient. The body properties are not changing so we can conclude that the entropy of the body itself is constant. The body is in a state of balance. However, entropy changes are occurring in the heat source and the heat sink. The entropy of the system as a whole is increasing so the system is not in a state of balance.

The second requirement is a test for hysteresis or "memory". The presence of these effects can be detected by approaching a data point P_1, T_1, X_1 from various directions. If all the dependent variables of state (the "B"s) are not the same, hysteresis or "memory" is present, and the body should not be considered as being in balance. Sufficient time should be allowed for applying this test to eliminate temporary effects. The effects of hysteresis are sometimes referred to as involving "internal variables of state". An unknown independent variables of state implies that the equation of state is also unknown and the thermodynamics as presented here is not applicable.

It is evident that gases and low viscosity liquids have little or no memory, for their properties are independent of their past history. Some high viscosity liquids are elastic. These store an elastic energy when they are flowing in shear; however, when flow stops a process of relaxation occurs in which the elastic energy dissipates itself and asymtotically approaches zero. These fluids have a fading memory. Their properties are simulated by a spring and dashpot in series (see chapters VII and VII). A viscoelastic fluid can be in a state of balance when it is flowing in a steady state of balance. When the flow stops there is a period of time, during which the elastic deformation relaxes, that the fluid is not in a state of balance. When relaxation is complete the fluid can again assume the state of balance.

Many solids have a memory or show a hysteresis which is effectively permanent, for solids cannot relax as easily as fluids. However, there are ways in which memories can be almost eliminated, such as by annealing or by oscillating with gradually decreasing amplitude around an apparent equilibrium point. Properties of bodies with memories are partially simulated by models which use friction blocks with the springs and dashpots. These will be described in Chapter VII.

The definition of the state of balance depends on whether the model of the system has two or more than two independent variables of state. Consider, for example, a weight hanging on a rubber band. Where the equation of state includes P, T, and X as the independent variables of state, the rubber body is in balance (and at equilibrium). However, if only two independent variables of state (P and T) are used in the model for which equilibrium is defined, the rubber body with its hanging weight is not at equilibrium. It will not be at equilibrium until the weights are removed and the rubber allowed to relax.

The selection of two or three independent variables of state for the definition of equilibrium is purely arbitrary. Either is valid. The selection of the particular definition of equilibrium (or balance) should depend on the use to be made of it in thermodynamic calculations, and how well the model of the system fits the actual system.

If a body is not in a state of balance within the restraints on the system it will do work to go to that state of balance. The maximum amount of work that the body will do to return to that state of balance within the specified restraints is the free energy.

Chapter VII
The Thermodynamics of Elastic Deformation

Pressure is a stress, and a change of volume is a strain. Thus stress and strain (although not usually identified as such) are inherent in thermodynamics even when limited to gases. The deformation of a gas (i.e., its compression or expansion) involves large thermal effects (and hence the name thermodynamics). Similar large thermal effects occur with the elastic deformation of rubberlike bodies, so the thermodynamics of rubber has been an historical area of interest in thermodynmaics. Treloar (1958) states that Kelvin was probably the first to consider the thermodynamic implications of the relationships between temperature and the elastic behavior of rubber. Recent concepts of the thermodynamics of rubber have been described by Treloar, by Flory (1953), and by many others.

Gibbs (1906) developed a broad thermodynamic treatment of the elastic deformation of solids. Interesting work has been done on the thermodynamics of the elastic deformation of metals, textile fibers, and other non-rubberlike polymers.

Uniform Stress

Stress is a tensor, however uniform stress can be described in terms of three normal forces commonly designated as p_{11}, p_{22}, p_{33}. These normal forces can also be considered as two normal forces imposed on a body under pressure. If X_3 and X_4 are two forces normal to each other which impose a uniform stress on a body under pressure P then $p_{11} = P$, $p_{22} = P + X_3$, and $p_{33} = P + X_4$. This implies the equation of state,

$$B = f(P, T, X_3, X_4)_m \qquad (1)$$

where B represents all the dependent variables of state including the strain (V, Y_3, Y_4), the entropy S, the internal energy E, etc. The subscript m designates that mass and composition are constant.

Solids can be stressed in such a way that the three normal stresses can be varied independently and yet remain uniform throughout a large area of the solid. One method of doing this is shown in Figure 1, which shows a sheet of material stretched in two directions. The third stress is the pressure. Only the center of the sheet is under uniform stress.

Uniform stress is more easily and practically obtained by using only one component of stress (other than pressure) which is independently variable. The equation of state is then,

$$B = f(P, T, X)_m \tag{2}$$

This may represent the equation of state for a rod under tension in which X represents the tension, or alternately a sheet in shear in which X represents the stress in shear. See Figures 2 and 3. These latter two examples of uniform stress are the most convenient examples to use for the development of the thermodynamics of stress and strain, and these along with the equation of state (2) will be continually referred to in the remainder of this book.

Most of the published work on the thermodynamics of rubber elasticity assumes that temperature T and tension X are the independent variables of state. This is equivalent to assuming that volume is constant. The resulting thermodynamic equations are the same as the conventional except that X and Y are substituted for P and V. This is a convenient approximation; for, with rubber the work associated with the change in volume during shear is usually 10^3 to 10^4 times less than the work done in shear (Treolar, 1958). Although convenient, this assumption can easily result in error and misunderstanding when the equations are applied to cases where the assumption is not valid. It is more precise and less misleading to note the constant volume assumption by using the V as a subscript on the partial differential equations for which the assumption is made.

The Elastic Energy and the Free Energy Functions

The elastic energy of deformation is the reversible work of deformation, but it is not a precise definition in thermodynamic terms. It is not precise until the thermal and pressure conditions under which the elastic work is done are precisely defined. Consider the following examples:

An elastic body is held at constant temperature in a bath and at constant pressure by the surrounding atmosphere. If this body is deformed elastically at so slow a rate that the frictional effects are negligible (i.e., there is no irreversible work) all the work done on the body is stored as elastic energy. By the first law this work done on the body must be equal to:

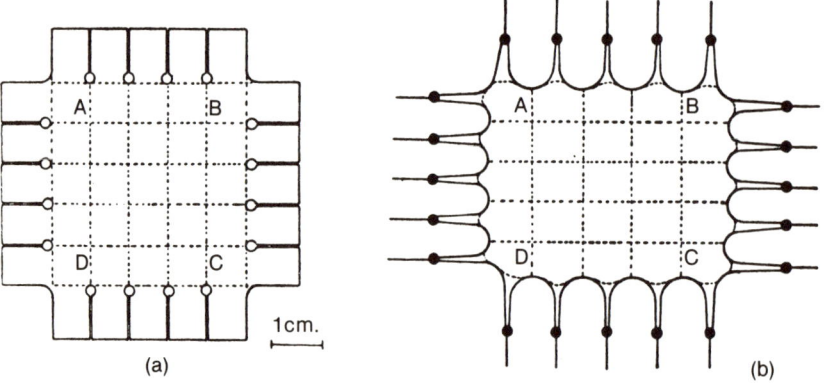

FIGURE 1. A method of stressing a sheet of rubber in two dimensions. (a) is in the unstrained state and (b) is in the strained state. The middle area ABCD undergoes pure homogeneous strain. From Treloar (1958).

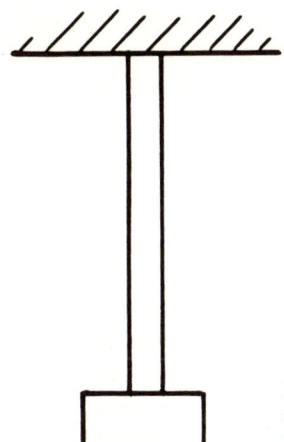

FIGURE 2. Subjecting a rod to uniform tension by suspending a weight by the rod. Note that the volume of the rod need not remain constant.

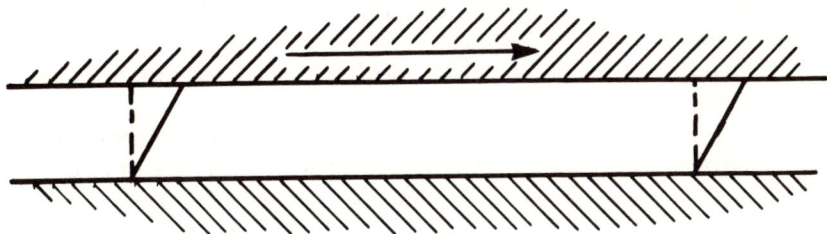

FIGURE 3. Subjecting a sheet to uniform shear. Note that the thermodynamic treatment does not require the volume to remain constant.

the change in internal energy (ΔE^{PT}), plus the heat absorbed from the bath ($-T \Delta S^{PT}$), plus the work done against the pressure of the surroundings from any change in volume ($P \Delta V^{PT}$). This is equal to a Δ Gibbs free energy function,

$$\Delta G^{PT} = \Delta E^{PT} - T \Delta S^{PT} + P \Delta V^{PT} \tag{3}$$

On the other hand if the conditions are constant V and T, the elastic energy of deformation (from the first law as above) is equal to a Δ Hemholtz free energy function.

$$\Delta A^{VT} = \Delta E^{VT} - T \Delta S^{VT} \tag{4}$$

If the conditions are adiabatic and constant pressure, then the elastic energy of deformation is a Δ change in enthalpy,

$$\Delta H^{SP} = \Delta E^{SP} + P \Delta V^{SP} \tag{5}$$

If the conditions are adiabatic and constant volume the elastic energy is equal to a change in internal energy,

$$\text{elastic energy} = \Delta E^{SV} \tag{6}$$

The above four relationships are not carefully delineated in the literature. Treloar (1958) and Green (1960) designate the elastic energy as a Helmholtz free energy function difference, and also the adiabatic elastic energy as an internal energy difference. Flory (1953) designates the elastic energy as a difference in the Gibbs free energy function as given above. Yang, Horne and Pound (1959) call the elastic energy of deformation of a metal a change in internal energy; however, in the published discussion of the paper several members of the audience state that the elastic energy should be a Gibbs free energy difference.

The term free energy usually applies specifically to a Gibbs function difference or a Helmholtz function difference. Free energy is a suitable term for all the expression (3, 4, 5, 6). The particular expression which applies depends on the system. Each has the traditional implications of free energy: a measure of the ability to do work, a measure of the tendency to change to another state at a lower free energy, a criterion of equilibrium, and a measure of escaping tendency. If there are ways in which work could be done by the strained body (other than the reversal of the original strain) such work may be done. Evidence of such a conversion of strain energy to other types of work is a confirmation of the concept that the strain energy is a free energy. The following are examples which confirm this concept:

1. Metal which has been cold worked and hence has internal stresses and strains will generally corrode more rapidly than metal which has not been cold worked. The stressed and unstressed metal form a galvanic cell which is electrically shorted. The passage of electricity consumes one of the electrodes (i.e., corrodes it). An example of this is the threads in steel pipe. They corrode much faster than the rest of the pipe because of the stresses and strains in the threaded section caused by the cutting of the threads. Stress corrosion is a major industrial problem.

2. When electrodes are prepared for use in the standard electrical cells (used as a reference in voltage measurements), care must be taken that strains or crystal imperfections are absent, for these strains and imperfections can affect the voltage of the cell (Lewis and Randall, p. 350, 1961).

3. "A Strained crystal behaves like an unstable polymorph in that it has a lower melting point, a higher solubility, and a higher vapor pressure." (Lindsay, p. 6, 1961) "A strained crystal or an imperfect crystal melts at a lower temperature than an unstrained or perfect one." (Nielsen, p. 63, 1962) See also Gibbs, (1906), Drigbaum (1967), Towle (1969), and Adams (1973).

4. Many rubbers form crystallites only on being stretched. These same crystallites disappear when the rubber is allowed to relax (Treloar, pp. 17, 235, 1958). The increase in the free energy function from the stretching causes the rubbers to seek another state with a lower free energy. In other words the crystallite state has a lower free energy than the amorphous stretched state but a higher free energy than the amorphous relaxed state.

5. Some chemical reactions take place only when a body is in the strained state (Casale and Porter, 1979). Such a chemical reaction

can be a scission of a chemical bond which generates free radicals which in turn react further in various ways. Examples of chemical reactions which take place in strained bodies are: (a) those which occur during the mastication of rubber — a normal step in the processing of rubber, (b) those which occur during the extrusion of plastics, and (c) the setting (polymerization) of certain adhesives which is initiated when they are subject to shear. Such an adhesive is Eastman 910 cement which is stated to be a cyanoacrylate ester.

For any spontaneous change to occur, it is necessary that the free energy be simultaneously lowered. Such a lowering of free energy occurs with: a spontaneous crystalization, a spontaneous chemical reaction, or any spontaneous change. In the above five examples the work of elastic deformation is the free energy which is spontaneously lowered.

In the thermodynamic treatment of elastic strain, the statement of the restraints on the system is sometimes omitted. The restraints on the system should always be stated. Their omission can lead to a wrong thermodynamic identification of the work of elastic deformation. The identification of the restraints shows which of the equations (3, 4, 5, or 6) is the free energy.

Rubberlike and Non-Rubberlike Elasticity

Just as there are ideal gases there are ideal elastic materials; however, instead of there being only one ideal material, as with a monatomic gas, there are two types of ideal elastic materials. In both types of ideal elasticity the frictional component of the elastic deformation is zero. It is best to compare these two types of ideal elasticity under two conditions: first under constant temperature and constant pressure, in which the work of elastic deformation is equal to ΔG^{PT} (equation 3), and second under constant entropy and constant volume in which the work of elastic deformation is equal to ΔE^{SV} (equation 5).

In the first type of elasticity all the work of elastic deformation is converted to molecular kinetic energy (E_k). When the conditions are constant temperature and pressure, ΔG^{PT} is equal to $-T \Delta S^{PT}$; and ΔE^{PT} and $P \Delta V^{PT}$ are equal to zero. All the heat derived from the work of elastic deformation is transferred to the surrounding constant temperature bath, causing a change in entropy. If, however, the conditions are constant entropy and constant volume, the heat generated cannot be transferred; it remains in the elastic body, causing the temperature to rise. This generated heat is equal to ΔE^{SV}. This first type of ideal elasticity is characteristic of rubber, and is referred to as rubberlike elasticity. The modulus of elasticity of rubberlike elasticity is relatively low.

In the second type of elasticity all the work of elastic deformation is converted to intermolecular potential energy (E_p). Under constant temperature, constant pressure no heat is transferred to a constant temperature bath. So, ΔG^{PT} is equal to ΔE^{PT}; and $-T \Delta S^{PT}$ and $P \Delta V^{PT}$ are equal to zero. Under constant entropy and constant volume conditions no molecular kinetic energy is generated so the temperature remains constant. The work of elastic deformation ΔE^{SV} is equal to ΔE_p. This type of ideal elasticity is characteristic of steel and many other solids where the modulus of elasticity is high. This is a non-rubberlike elasticity.

Rubberlike Elasticity

Some of the characteristics of rubberlike elasticity can be demonstrated with a rubber band using the experiments described by John Gough in 1805 (Treloar, pp. 10, 38, 1958). First stretch a rubber band while in contact with the upper lip. A sensation of warmth will be felt because of the heat generated on stretching. Then allow the rubber band to contract, while still pulling on it so that it does external work in contracting. A sensation of coolness will be felt because of the absorption of heat on contraction. The lip is used for these tests because of its sensitivity to temperature.

Second, take a rubber band and hang a weight on it of sufficient size to stretch the band moderately. Then warm the rubber band slightly with a lighted match. The rubber band will shorten, lifting the weight. When the band is allowed to cool the weight will be lowered. These two experiments demonstrate the close relation between the thermal and the mechanical properties of rubberlike materials.

There are very similar behaviors between the ideal rubber and the ideal gas. This similarity is shown in the following three ways:

(a) All the work done during the compression of an ideal gas is converted to heat (molecular kinetic energy); likewise, the work done by the gas during an expansion against a restraining pressure is done by the consumption of heat. When the ideal gas is held at constant temperature this relationship is given by,

$$\int P^T \, dV = -T \, \Delta S^T \qquad (7)*$$

With an ideal rubber all the work done on the rubber during an elastic deformation is converted to heat. With elastic recovery all the work done by the rubber is done at the expense of heat. When the ideal rubber body is held at constant temperature this relationship is given by,

$$\int X^{TV} \, dY = -T \, \Delta S^{TV} \qquad (8)*$$

(b) The equation of state for an ideal gas is $PV = nkT$. From this it is evident that at constant volume the pressure is proportional to the absolute temperature or,

$$P^V \propto T \qquad (9)*$$

With an ideal rubber the magnitude of the force of deformation (in extension, compression, or shear) is proportional to the absolute temperature

$$|X^{YV}| \propto T \qquad (10)*$$

where | | indicates absolute value. It is used to eliminate the effects of the difference in the sign of compression and extension.

(c) One can state the entropy changes in the two above examples in terms of probability. The entropy change of an ideal gas in going from volume (1) to volume (2) is $k \ln W$, where W is the probability that the molecules in volume (1) will all be in volume (2) a part of volume (1) at any given instant by chance and chance alone.

The entropy change of an ideal rubber is also $k \ln W$, where W is the probability that the molecule with a conformation of state (1) will have a conformation of state (2) by chance and chance alone. The conformation of the rubberlike molecules will be discussed later.

Spring and Dashpot Models

Real rubbers are viscous as well as elastic (i.e., they are viscoelastic). Their deformation involves irreversible (viscous) as well as reversible (elastic) work.

Viscoelastic materials are frequently represented by spring and dashpot models. The spring represents the capability to convert work into a reversibly available form. The dashpot represents the capability to dissipate the work into an irreversible form of heat. Figure 4a shows a spring and dashpot in parallel — the simplest model for a viscoelastic solid. Friction blocks in models represent the ability of a viscoelastic body to have yield values or a memory (see Figure 4b). Notice that while model 4a will return to zero displacement and zero elastic energy upon being allowed to relax, the model in Figure 4b will not. Some elastic energy remains stored in the spring, hence, the model in Figure 4b has a memory. If it is displaced in a cyclic way it will always show hysteresis.

These models can be made more complex by adding more springs and dashpots each with a different constant (see Figure 5). When the displacement of the spring is porportional to the force exerted by the spring and the rate of displacement of the dashpot is proportional to the force of resistance exerted by the dashpot, the models are called linear. The linear models are particularly adaptable to mathematical manipulation so these are thoroughly described in the literature. These linear models have been particularly useful in representing the behavior of polymers when subjected to small sinusoidal displacements over a range of frequencies. The response of polymers and polymer solutions at various frequencies and temperatures has a molecular interpretation which is useful. However, the linear models are not adequate for large elastic deformations.

In terms of thermodynamics these models are also inadequate as they do not distinguish between the two types of elasticity. They also do not distinguish between adiabatic and isothermal displacements.

On the other hand the spring and dashpot models do illustrate an important aspect of thermodynamics — the separability of the reversible and irreversible processes. The action of the dashpots represent the irreversible

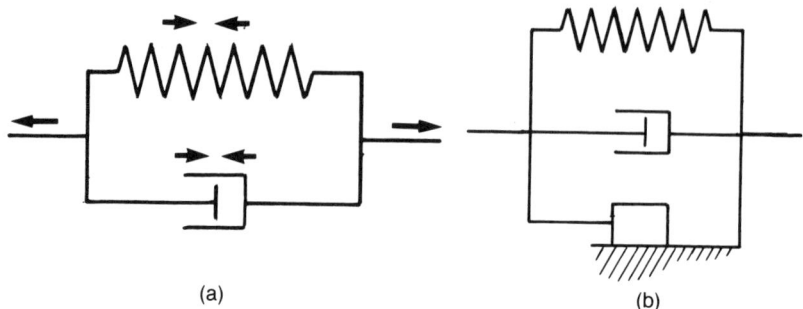

FIGURE 4. Simple models of viscoelastic solids. (a) is the Voight model of a spring and dashpot in parallel. (b) adds a friction block representing a solid with a yield value and hysteresis.

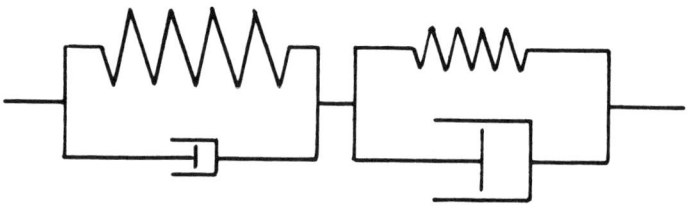

FIGURE 5. Models of solids with a more complex linear response are obtained by linking spring and dashpot models together in series. The various springs and dashpots have different constants.

components of the action and it is completely separable from the actions of the spring which represent the reversible component of the action. The spring and dashpot are separate on the diagram, and the two actions are likewise thermodynamically separable. This aspect will be discused further in Chapter VIII.

The dashpot component in the deformation of rubbers can be relatively large or small. Commercial rubbers are formulated to have a large or small irreversible action depending on the desired use. Rubbers with a small irreversible component are called "live" and balls made with them will bounce almost as high as the height from which they are dropped. Rubbers made with a relatively large irreversible action are called "dead" and will bounce very poorly if at all. They are useful as mounts for machinery to minimize the effects of vibration.

A Molecular Model for Rubberlike Elasticity

The generally accepted molecular model for rubber is a loose network of long carbon chains fastened together at random points called cross-links. Between the cross-links the chains are in violent thermal motion — moving in all possible ways and taking all possible shapes. Under the actions and the restraints of the attached chains the cross-links move about average positions.

If a tension is imposed on rubber the average positions of the cross-links are displaced, and the thermal motion of the chains exerts a force tending to bring the cross-links back to their average positions. Just as the presssure exerted by the impact of the molecules of an (ideal) gas is proportional to the absolute temperature, the force exerted by the thermal motions of the chains is proportional to the absolute temperature (for ideal rubbers). Since real rubbers expand in volume with an increase in temperature it is best to use a constant extension ratio rather than constant length for the experimental study of the relationship between tension and temperature (Treloar 1958).

When the extension of rubber is large a portion of the carbon chains becomes completely extended. As the extension increases this proportion of completely extended chains increases and the tension force rises rapidly until further extension causes a tearing of the rubber.

When rubber is cooled, a temperature is reached at which the carbon chains loose their mobility, and, hence the rubberlike elasticity is lost. The rubber is then stated to be in the glassy state. There is no large release of heat as when liquids solidify; however, there is a change in the specific heat. Other properties such as specific volume behave in a similar way. This is called a second order transition. The difference between first and second order transitions is shown by plots of enthalpy versus temperature. First order transitions show a vertical rise in enthalpy at melting and boiling points. Second order transitions show only a change in slope at the glassy point when the enthalpy is plotted against temperature.

Some rubbers form crystallites under stretching or cooling. These crystallites may be shorter in length than the carbon chains. Portions of separate carbon chains crystallize together leaving other portions of the chains to take random positions between the crystallites. The crystallites which form on stretching are primarily orientated in the direction of elongation. The crystallites which form on cooling have a random orientation.

Real Rubbers

Real rubbers show hysteresis when they are stretched and allowed to recover on a cyclic basis. On some rubbers this is quite small is the cycling is sufficiently slow. However, when hysteresis is appreciable the equations of state do not apply and the equations presented here do not apply, or they must be regarded as an approximation.

Real rubbers increase in volume when stretched or when heated. Treloar states that the constant volume assumption is reasonable when the extensions are over twenty percent. For convenience the volume will usually be assumed constant; however, when this is done it will always be noted on the equations used.

A Comparison Between the Thermodynamics of Rubbers and Gases

There is a very close analogy between the thermodynamic behavior of gases and that of viscoelastic bodies. This is especially evident in the temperature changes which occur when a body is first strained adiabatically and then allowed to relax without doing external work. These processes of adiabatic strain followed by relaxation without doing external work will be reviewed first in gases and then discussed and explained in rubberlike polymers.

Consider first an ideal gas whose molecules do not attract or repel each other. Figure 6 shows the calculated temperature changes in such a gas when it is first compressed adiabatically from one atmosphere pressure to 15 atmospheres pressure, and then allowed to relax to one atmopshere pressure without doing external work. When such an ideal gas is compressed adiabatically all the work done on the gas is converted to translational kinetic energy. This is evident from the facts that the only form of internal energy for such a gas is translational kinetic energy and since the compression is adiabatic all work done on the gas must remain in the gas — as internal energy. That is,

$$\int {}^S P\, dV = C_V (T_2 - T_1) = (3/2) Nk (T_2 - T_1) \qquad (11)*$$

The superscript "S" on "P" indicates a constant entropy integration.

When this ideal gas is allowed to relax without doing external work (i.e., expand into a vacuum) the kinetic energy of the molecules remains the same so there is no temperature change. Such a relaxation is at constant internal energy and at constant temperature.

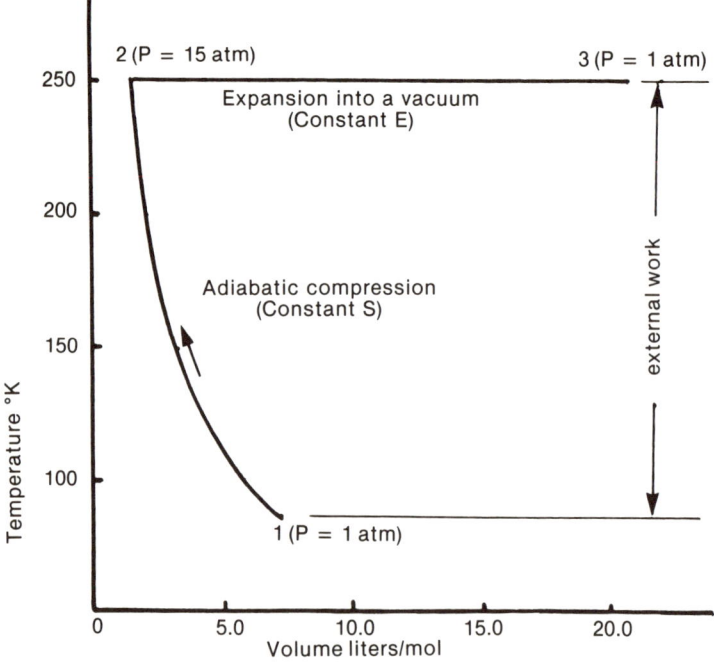

FIGURE 6. Adiabatic compression followed by expansion into a vacuum for an ideal monatomic gas.

THE THERMODYNAMICS OF ELASTIC DEFORMATION

The simplest real gases are the heavy monatomic gases which have been previously described in Chapter IV. At moderate temperatures and pressures their molecules follow the laws of classical mechanics, and it is not necessary to use quantum mechanics to interpret their behavior. The molecules attract and repel each other, and the internal energy is the sum of the kinetic and the potential energy of the molecules. This can be written as,

$$E = E_k + E_p \qquad (12)**$$

Argon is such a gas and Figure 7 shows the temperature changes which occur when it is first compressed adiabatically and then allowed to relax by expanding into a vacuum. The paths shown were calculated from Gosman's (1969) thermodynamic tables for argon.

When the argon gas is compressed adiabatically over the same pressure range as the ideal gas the temperature increases more than with the ideal gas. This is because the attraction of the molecules aids the external force in the compression of argon. This results in internal work which is converted to molecular kinetic energy in addition to the molecular kinetic energy generated by the external work. When the argon is allowed to expand into a vacuum until it returns to the original pressure, the thermal motion of the molecules does work against the intermolecular forces. This converts molecular kinetic energy into intermolecular potential energy. Although the total internal energy remains the same the conversion of kinetic energy into potential energy decreases the temperature.

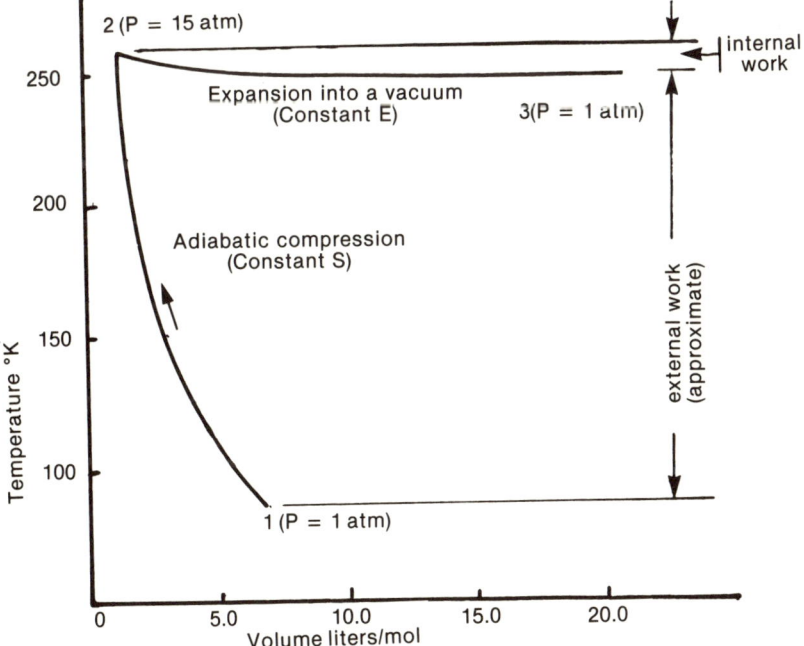

FIGURE 7. Adiabatic compression followed by expansion into a vacuum for argon gas.

Consider a body of argon gas which is allowed to expand but kept at constant temperature. Since the temperature is constant the molecular kinetic energy is constant. Hence all the heat absorbed during this expansion is converted to intermolecular potential energy. All the increase in internal energy is an increase in intermolecular potential energy. This quantity is known as the internal work and is equal to,

$$\int (\partial E/\partial V)_T \, dV = \int (T(\partial P/\partial T)_V - P) \, dV \qquad (13)$$

where $(\partial E/\partial V)_T$ is known as the internal pressure and is also written as Pi.

For the heavy monatomic gases during adiabatic compression the loss in intermolecular potential energy is,

$$\Delta E_p^S = \int {}^S Pi \, dV = \int {}^S (\partial E/\partial V)_T \, dV \qquad (14)**$$

where the increase in molecular kinetic energy is,

$$\Delta E_k^S = - \int {}^S (P + Pi) \, dV = (3/2) Nk (T_2 - T_1) \qquad (15)**$$

For the expansion of argon into a vacuum,

$$E = \text{constant}, \quad \Delta E_k^E + \Delta E_p^E = 0 \qquad (16)**$$

hence from equations (15) and (16),

$$\Delta E_p^E = -E_k^E = \int {}^E Pi \, dV = (3/2) Nk (T_2 - T_3) \qquad (17)**$$

If we could have an ideal rubber we could compare its behavior to the above ideal gas. It would be perfectly analogous. On stretching adiabatically its temperature would rise. If it were allowed to snap back without doing external work its temperature would not change.

Real rubbers are analogous to real gases. When they are stretched their temperature will rise; however, on allowing them to snap back without doing external work their temperature will fall. The temperature drop is relatively much more than the temperature drop for argon shown in Figure 7. This behavior can be checked with a rubber band by comparing its temperatures with the upper lip — before stretching, when stretched, and after allowing the rubber to snap back without doing external work.

These temperature changes are associated with an internal work term analogous to that for gases,

$$\int (\partial E/\partial Y)_{TV} dY = \int (T(\partial X/\partial T)_{YV} - X) \, dY = \int Xi \, dY \qquad (18)$$

where $(\partial E/\partial Y)_{TV}$ or Xi is called the internal tension.

The internal tension is best construed as a lowering in the pull or tension in a rubber by various intermolecular forces which tend to help align the molecular chains with each other. The release of heat may or may not be accompanied by the formation of crystallites. The formation of crystallites in strained rubber is an external confirmation of the partial alignment of the molecular chains.

Temperature changes during a rapid stretching and then relaxation have been measured by Good (1971).† He used a tape with an acrylic

---

†However the explanation given here is not due to Good.

copolymer based adhesive. A very small thermocouple (0.001 inch wires) was placed in the adhesive, and then the tape was peeled from a surface by pulling it with an Instron. The rapid temperature changes were recorded with a Honeywell Visicorder. An example of the results he obtained is shown in Figure 8. The temperature first rose rapidly to a sharp peak with a remarkable increase in temperature of 12.5 °C (ΔT_1). This temperature rise is caused by the release of heat equal to the sum of the external and internal work. After the peak there is a rapid decrease in temperature of 9.7 °C (ΔT_3). This is caused by the internal work done against the intermolecular forces by the thermal motion of the polymer chains. The work done by this thermal motion causes a loss in the kinetic energy of these chains, lowering their temperature. Since the internal work is always completely reversible no matter what its sign, the drop in temperature ΔT_3 is hence also a measure of the internal work done in the stretching of the polymer. During the extension small crystallite may have formed in the acrylic copolymer, and the release of their heat of crystallization during their formation and the absorption of that heat during their melting may in part account for the high peak. This does not change the interpretation, for such a heat of crystallization is a proper component of the internal work.

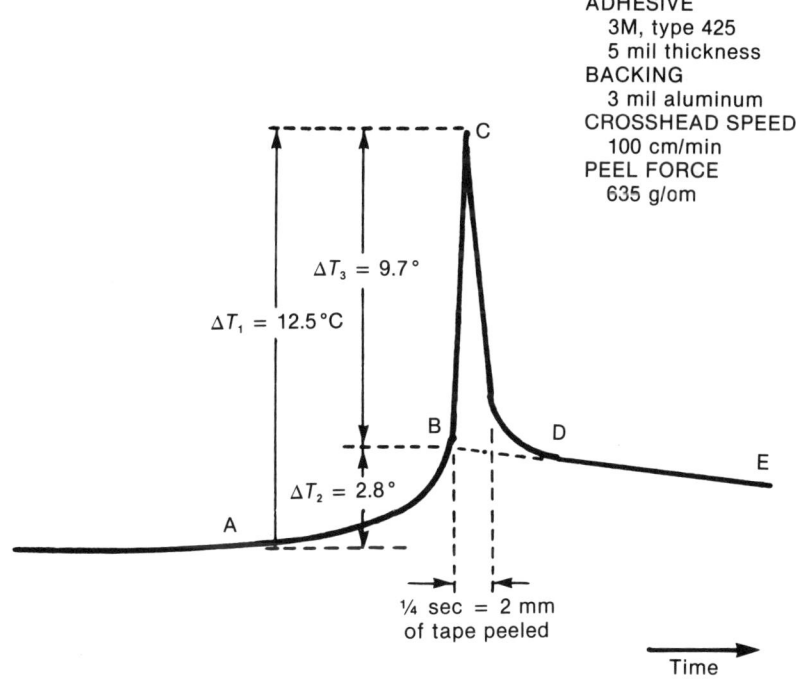

FIGURE 8. The temperature changes which occur in the adhesive of a tape as it is pulled from a surface. The total temperature rise (ΔT_1) is much greater than the temperature rise calculated from the work done in peeling the tape from the surface (ΔT_2). From Good (1977), redrawn by Hull.

The residual temperature ΔT_2 is a measure of the external work done on the polymer during its extension. This is the sum of the work done against the viscous flow, plus the work done against the elastic deformation. The latter was made irreversible by the relaxation of the polymer. There is no way of telling what proportion of the external work was initially irreversible; however, it is not important in the interpretation of the meaning of the three temperature differences.

Figure 9 shows the residual temperature rise (ΔT_2) for various rates of separation of the tape. The solid line shows the calculated temperature rise from the work done during the separation of the tape. Their agreement over a range of speeds confirms that ΔT_2 is a measure of the total external work done during the separation.

The whole process diagramed in Figure 8 takes place so rapidly that it is essentially adiabatic. There is a slight loss of heat, which is indicated by the rate of change of temperature after the initial fall in temperature. A correction for this heat loss is made by extending the near horizontal line to the time of initial temperature rise (as shown by the dotted line in Figure 8).

In summary, —

The rise in temperature ΔT_1 is a measure of the heat released during the stretching of the polymer. This is the heat equivalent of the sum of the external work done on the polymer (reversible and irreversible) and the internal work done by the intermolecular forces to help move the polymer chains to their strained (i.e., less probable) positions.

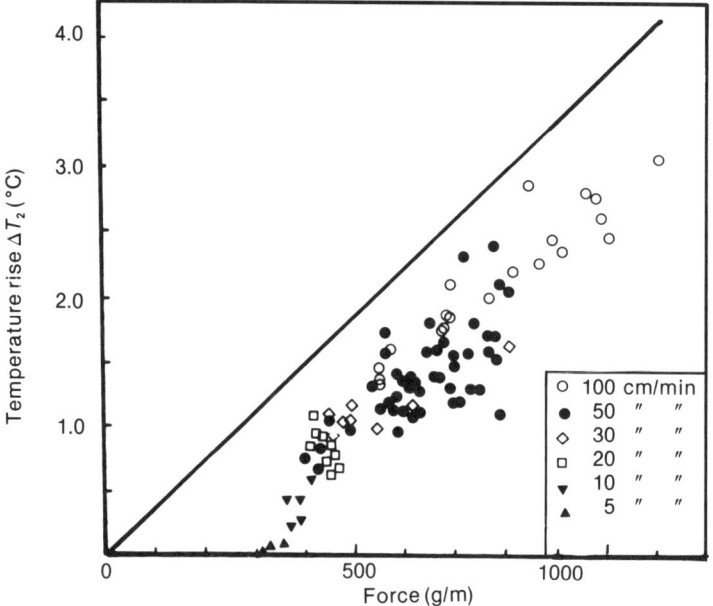

FIGURE 9. A plot of the residual temperature rise (ΔT_2 in Figure 6) for various rates of speed for tape removal. Note that the residual temperature rise is just below that calculated from the external work done on the tape adhesive (represented by the straight line). From Good (1977) and redrawn by Hull.

The residual temperature rise ΔT_2 is a measure of the total irreversible work which results from the whole cycle of stretching and relaxation. It is equal to the total external work done during the stretching of the polymer. Although the external work has a reversible component at the time of stretching, this component becomes irreversible during the relaxation.

The fall in temperature after the peak, ΔT_3, is a measure of the internal absorption of the heat during the relaxation of the stress. This is equal to the work done by the thermal motion of the molecules against the intermolecular forces as the polymer chains are moved to a more random position. If crystallites are formed during the stretching, their heat of fusion is included in ΔT_3.

Energy Flow During Elastic Deformation

There are four types of energy (or work) involved in the elastic deformation of a body. During a bodies elastic deformation the energy can be visualized as being transformed from one of these types to another by work processes. These four elements of the work processes are shown in the diagram in Figure 10, which is a variation of the standard spring and dashpot diagram. These four types of work are,

(a) *The external work* — that is work done *by* the body on external forces (positive work), or work done *on* the body by external forces (negative work). The forces which do this work are completely reversible and they are a part of the system or the environment in which the body is placed. These external forces include: pressure, tension, compression, shear, etc. These various forces may all act at the same time or act individually but their action in the figure is indicated by the arrows labeled (a).

(b) *The work done by or against the thermal motion* of the molecules or polymer chains. If the work is done adiabatically against the thermal motion, it results in an increase in the kinetic energy of the molecules (i.e., an increase in temperature). If the work is done at constant temperature, it results in the transfer of heat to a heat sink. The capability for this work is indicated in Figure 10 by the molecular kinetic energy spring (b). Work done against this spring is completely reversible; however, it is reversible only because the polymer chains are forced to less probable positions. If there occurs an internal relaxation process, by means of which the polymer chains return to more probable positions, the reversibility is lost; the heat generated becomes irreversible. This molecular kinetic energy spring is the primary source of rubberlike elasticity.

(c) *The work done by or against the intermolecular forces.* This capability is represented in Figure 10 by the molecular potential energy spring (c). In thermodynamic terms this is work done by or against the internal pressure (Pi) or the internal tension (Xi) or its equivalent for other types of deformation. It is completely reversible and never subject to any internal relaxation process as the

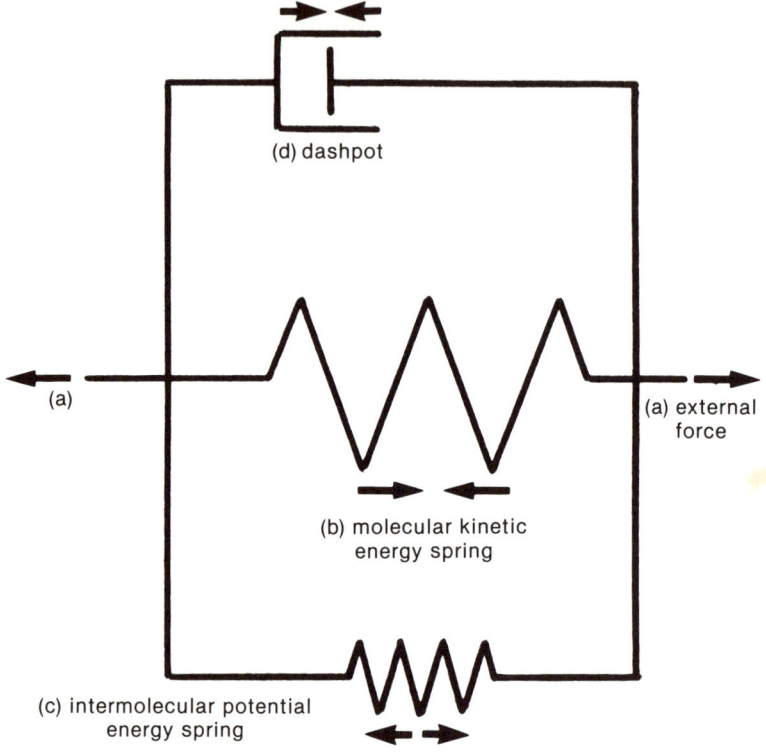

FIGURE 10. Dashpot and spring model of viscoelastic solids showing separate molecular kinetic energy and intermolecular potential energy springs. With real rubberlike materials the kinetic energy spring dominates and the intermolecular potential energy spring acts in opposition to the kinetic energy spring. With an ideal rubber the actions of the dashpot and the potential energy spring are negligible. With ideal nonrubberlike solids the potential energy spring dominates and the actions of the other two elements are neglible.

kinetic energy spring may be. In ideal gases and ideal rubbers the molecular potential energy spring is non-existant. In real gases and most rubbers the potential energy spring acts to aid the external forces in opposing the kinetic energy spring. In hard non-rubberlike bodies the potential energy spring opposes the action of the external forces. This spring is the source of non-rubberlike elasticity.

(d) *The work done in causing viscous flow.* This capability is indicated in Figure 10 by the dashpot (d). This work is completely converted to heat just as in the entropy spring (b) but it differs in that it is completely irreversible. The viscous work is commonly the result of the action of the outside forces (a); however, it can also be the result of the action of (b) or (c) above during a relaxation process.

It is interesting to compare the various types of elasticity by another type of plot. Figure 11 shows the ratio of the heat released during elastic

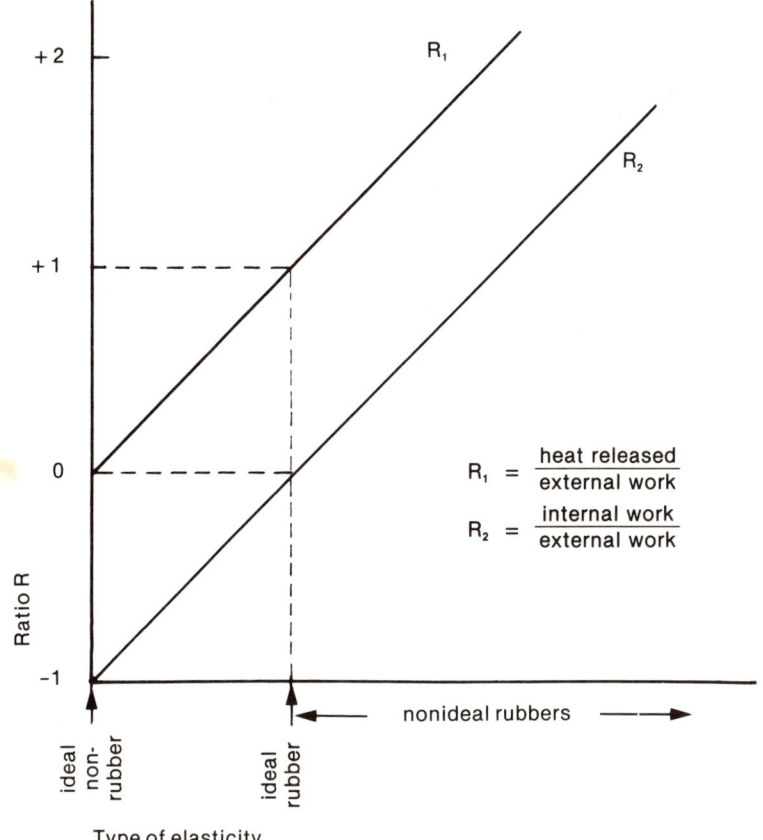

FIGURE 11. A diagram of the relative amount of heat released and the relative amount of internal work for the various kinds of elastic deformation. Note that the internal work is zero for the ideal rubber, is negative for the ideal nonrubbers, but is positive for real nonideal rubbers.

deformation, divided by the external work (R_1). This is plotted against the type of elastic deformation. No heat is released with the ideal non-rubber as all the external work is done against the intermolecular forces. With the ideal rubber the heat released is equal to the external work as there are no intermolecular forces and all the work is done against the thermal motion of the molecules. With non-ideal rubber and real gases the intermolecular forces aid the external forces in forcing the molecules or polymer chains to a less random position; so, the heat released is equal to the sum of the external and internal work.

Figure 11 also shows a plot of the internal work divided by the external work (R_2) for the same materials. This is zero for the ideal rubber, positive for the non-ideal rubbers, and negative for non-rubbers. The irreversible processes as represented by the dashpot in Figure 10 are considered to be negligible in Figure 11.

Equilibrium with Three Independent Variables of State

It has been shown that the elastic energy at constant T and P is a ΔG^{PT}. The criterion of equilibrium at a given P and T is a minimum G (which also can be stated to be a minimum ΔG^{PT}). Under this definition there can be no stored elastic energy at equilibrium.

Consider now a model with three independent variables of state — a weight hanging on a rubber body at constant P and T. If any work is done on that rubber body to throw it out of equilibrium, this will result in a change in the internal energy ΔE^{PTX}, a flow of that from the body equal to $-T \Delta S^{PTX}$, plus work done against the atmosphere $P \Delta V^{PTX}$, plus any work done in lifting the weight $X \Delta Y^{PTX}$. This is equal to,

$$\Delta GG^{PTX} = \Delta E^{PTX} - T \Delta S^{PTX} + P \Delta V^{PTX} + X \Delta Y^{PTX} \qquad (19)$$

Since ΔGG^{PTX} is an amount of work done to put the system out of equilibrium, it must also be at a minimum for the system to be at equilibrium. It is thus a free energy, and must be added to the list of free energies in equations (3) through (6) given at the beginning of this chapter.

In Chapter VI, six conditions for equilibrium are listed for two independent variables of state. For three independent variables of state there are twelve conditions for equilibrium; however, in order to state these new conditions, new thermodynamic functions need to be defined. These are:

$$GG = E - TS + PV + XY \qquad (20)$$

$$HH = E + PV + XY \qquad (21)$$

$$HA = E - TS + XY \qquad (22)$$

$$AA = E + XY \qquad (23)$$

The six conditions for equilibrium at constant X are:

for any given T, P and X, that GG is a minimum (23)

for any given S, P and X, that HH is a minimum (24)

for any given S, V and X, that AA is a minimum (25)

for any given AA, V and X, that S is a maximum (26)

for any given T, V and X, that HA is a minimum (27)

for any given HH, P and X, that S is a maximum (28)

The six conditions for equilibrium at constant Y are directly analogous to the conditions for two independent variables of state given in Chapter VI. They are:

for any given E, V and Y, that S is a maximum (29)

for any given S, P and Y, that H is a minimum (30)

for any given T, P and Y, that G is a minimum (31)

for any given H, P and Y, that S is a maximum (32)

for any given S, V and Y, that E is a minimum (33)

for any given T, V and Y, that A is a minimum (34)

Although all of these conditions for equilibrium are not expected to be useful, their listing points out their relationship to an arbitrary system which is assumed to represent a real system. Also these conditions are not equivalent — in particular those systems with three independent variables of state versus those with only two independent variables of state.

Partial Differential Equations for Three Variable Systems

The introduction of the third independent variable of state changes the partial differential equations. While some of the new equations differ from the standard equations by only minor ways, and these equations can sometimes be inferred by analogy, it is best to develop the new equations from the beginning by means of the standard methods.

From the first law the expression for the differential internal energy of a closed system is,

$$dE = T\,dS - P\,dV - X\,dY \qquad (35)$$

By differentiation of the defining equations for H, A, G, GG, HH, HA, and AA (see Chapter V and equations 20, 21, 22, 23 above) and substracting equation (35) we obtain:

$$dH = T\,dS + V\,dP - X\,dY \qquad (36)$$
$$dA = -S\,dT - P\,dV - X\,dY \qquad (37)$$
$$dG = -S\,dT + V\,dP - X\,dY \qquad (38)$$
$$dGG = -S\,dT + V\,dP + Y\,dX \qquad (39)$$
$$dHH = T\,dS + V\,dP + Y\,dX \qquad (40)$$
$$dHA = -S\,dT + Y\,dX - P\,dV \qquad (41)$$
$$dAA = T\,dS - P\,dV + Y\,dX \qquad (42)$$

From equations (35) through (42) the following relationships are obtained by holding two variables constant:

$$(\partial H/\partial Y)_{PS} = (\partial A/\partial Y)_{TV} = (\partial G/\partial Y)_{TP} = (\partial E/\partial Y)_{SV} = -X \quad (43)$$
$$(\partial HH/\partial X)_{PS} = (\partial GG/\partial X)_{PT} = (\partial HA/\partial X)_{TV} = (\partial HA/\partial X)_{SV} = Y \quad (44)$$
$$(\partial A/\partial T)_{VY} = (\partial G/\partial T)_{PY} = (\partial GG/\partial T)_{PX} = (\partial HA/\partial T)_{XV} = -S \quad (45)$$
$$(\partial E/\partial S)_{VY} = (\partial HH/\partial S)_{PX} = (\partial H/\partial S)_{PY} = (\partial AA/\partial S)_{VX} = T \quad (46)$$
$$(\partial H/\partial P)_{SY} = (\partial HH/\partial P)_{XS} = (\partial G/\partial P)_{TY} = (\partial GG/\partial P)_{TX} = V \quad (47)$$
$$(\partial E/\partial V)_{SY} = (\partial A/\partial V)_{TY} = (\partial HA/\partial V)_{XT} = (\partial AA/\partial V)_{XS} = -P \quad (48)$$

By differentiating the relationships in (43) through (48) a second time, the analogs of the Maxwell relations are obtained:

$$(\partial^2 H/\partial Y\partial S)_P = (\partial^2 H/\partial S\partial Y)_P = -(\partial X/\partial S)_{YP} = (\partial T/\partial Y)_{SP} \quad (49)$$
$$(\partial^2 H/\partial Y\partial P)_S = -(\partial X/\partial P)_{YS} = (\partial V/\partial Y)_{PS} \quad (50)$$
$$-(\partial^2 H/\partial S\partial P)_Y = (\partial T/\partial P)_{SY} = (\partial V/\partial S)_{PY} \quad (51)\#$$
$$-(\partial^2 A/\partial Y\partial T)_V = (\partial X/\partial T)_{YV} = (\partial S/\partial Y)_{TV} \quad (52)$$

$$-(\partial^2 A/\partial Y \partial V)_T = (\partial X/\partial V)_{YT} = (\partial P/\partial Y)_{VT} \tag{53}$$

$$-(\partial^2 A/\partial T \partial V)_Y = (\partial S/\partial V)_{TY} = (\partial P/\partial T)_{VY} \tag{54}$$

$$-(\partial^2 G/\partial Y \partial T)_P = (\partial X/\partial T)_{YP} = (\partial S/\partial Y)_{TP} \tag{55}$$

$$(\partial^2 G/\partial Y \partial P)_T = -(\partial X/\partial P)_{YT} = (\partial V/\partial Y)_{PT} \tag{56}$$

$$(\partial^2 G/\partial T \partial P)_Y = -(\partial S/\partial P)_{TY} = (\partial V/\partial T)_{PY} \tag{57}\#$$

$$(\partial^2 E/\partial Y \partial S)_V = -(\partial X/\partial S)_{YV} = (\partial T/\partial Y)_{SV} \tag{58}$$

$$-(\partial^2 E/\partial Y \partial V)_S = (\partial X/\partial V)_{YS} = (\partial P/\partial Y)_{VS} \tag{59}$$

$$(\partial^2 E/\partial S \partial V)_Y = (\partial T/\partial V)_{SY} = -(\partial P/\partial S)_{VY} \tag{60}\#$$

$$(\partial^2 HH/\partial X \partial S)_P = (\partial Y/\partial S)_{XP} = (\partial T/\partial X)_{SP} \tag{61}$$

$$(\partial^2 HH/\partial X \partial P)_S = (\partial Y/\partial P)_{XS} = (\partial V/\partial X)_{PS} \tag{62}$$

$$(\partial^2 HH/\partial S \partial P)_X = (\partial T/\partial P)_{SX} = (\partial V/\partial S)_{PX} \tag{63}\#$$

$$(\partial^2 GG/\partial X \partial T)_P = (\partial Y/\partial T)_{XP} = -(\partial S/\partial X)_{TP} \tag{64}$$

$$(\partial^2 GG/\partial X \partial P)_T = (\partial Y/\partial P)_{XT} = (\partial V/\partial X)_{PT} \tag{65}$$

$$(\partial^2 GG/\partial T \partial P)_X = -(\partial S/\partial P)_{TX} = (\partial V/\partial T)_{PX} \tag{66}\#$$

$$(\partial^2 HA/\partial X \partial T)_V = (\partial Y/\partial T)_{XV} = -(\partial S/\partial X)_{TV} \tag{67}$$

$$(\partial^2 HA/\partial X \partial V)_T = (\partial Y/\partial V)_{XT} = -(\partial P/\partial X)_{VT} \tag{68}$$

$$-(\partial^2 HA/\partial T \partial V)_X = (\partial S/\partial V)_{TX} = (\partial P/\partial T)_{VX} \tag{69}\#$$

$$(\partial^2 AA/\partial X \partial S)_V = (\partial T/\partial S)_{XV} = (\partial Y/\partial X)_{SV} \tag{70}$$

$$(\partial^2 AA/\partial X \partial V)_S = (\partial Y/\partial V)_{XS} = -(\partial P/\partial X)_{VS} \tag{71}$$

$$(\partial^2 AA/\partial S \partial V)_X = (\partial T/\partial V)_{SX} = -(\partial P/\partial S)_{VX} \tag{72}\#$$

The eight equations marked with the (#) are identical with the four Maxwell relations listed in Lewis and Randall (1961, p. 665) except that an additional variable is held constant. There are eight equations (with #) since four have X held constant and four have Y held constant. There are similar relationships with other equations. For example, equation (49) differs only from equation (58) in that in the first P is held constant and in the second V is held constant.

There are other thermodynamics relations which can be developed, and some of these are illustrated in the discussion of the Δ functions which follows.

The Δ Functions

Some examples of the meanings and uses of the Δ functions follow. The superscripts introduced earlier are particularly useful as a help in the interpretation of their meaning and as a guide in the development of the correct thermodynamic relationships.

It has been shown that ΔG^{PT} is equal to the stored elastic energy. This is equal to,

THE THERMODYNAMICS OF ELASTIC DEFORMATION 69

$$\Delta G^{PT} = (G - G_o)^{PT} \tag{73}$$

where G_o is the unstrained rubber at temperature T and pressure P. Partial differential equations can be developed for the rate of change of the elastic energy with the various variables such at T, P, X, and Y. In doing this the reference state G_o changes to that of the unstrained state at the pressure and temperature involved.

By taking the partial derivative of the equation for ΔG^{PT} with respect to Y keeping P and T constant,

$$\Delta G^{PT} = \Delta E^{PT} - T \Delta S^{PT} + P \Delta V^{PT} \tag{74}$$

we obtain,

$$(\partial \Delta G^{PT}/\partial Y)_{PT} = (\partial \Delta E^{PT}/\partial Y)_{PT} - T(\partial \Delta S/\partial Y)_{PT} + P(\partial \Delta V^{PT}/\partial Y)_{PT} \tag{75}$$

The derivative with respect to temperature may also be taken, keeping P and Y constant. This gives,

$$(\partial \Delta G^{PT}/\partial T)_{PY} = (\partial \Delta E^{PT}/\partial T)_{PY} - \Delta S^{PT} - T(\partial \Delta S^{PT}/\partial T)_{PY} + P(\partial \Delta V^{PT}/\partial T)_{PY} \tag{76}$$

If the extension Y is kept constant we can write from the first law,

$$\Delta E^{PTY} - T \Delta S^{PTY} + P \Delta V^{PTY} = 0 \tag{77}$$

If the tension X is kept constant we obtain from the first law,

$$\Delta E^{PTX} - T \Delta S^{PTX} + P \Delta V^{PTX} + X \Delta Y^{PTX} = 0 \tag{78}$$

Thermodynamics of Swelling of Polymers Under Stress

The Clapeyron equation is developed from the Maxwell relationship,

$$(\partial P/\partial T)_V = (\partial S/\partial V)_T = \Delta H^{PT}/(T \Delta V^{PT}) = dP/dT \tag{79}$$

It applies to the relationship between the variation of vapor pressure with temperature and its liquid or solid phase. ΔH^{PT} is the heat of evaporation and ΔV^{PT} is the change in volume on evaporation. The constant volume restriction on $(\partial P/\partial T)_V$ can be dropped because volume does not affect the vapor pressure. The Clapeyron applies to the vapor pressure of a solvent in a swollen rubber where ΔH^{PT} is the heat of evaporation, per mole of the solvent, from that swollen polymer and ΔV^{PT} is the change in volume, per mole of solvent, on evaporation from that swollen polymer. This provides a whole series of curves for change of vapor pressure — one for each concentration of solvent in the polymer.

From equations (54) and (69) the equation can be modified to apply to swollen polymers under constant stress or under constant strain. We then obtain,

$$dP/dT = \Delta H^{PTX}/(T \Delta V^{PTX}) \tag{80}$$

where ΔH^{PTX} is the heat of evaporation, per mole of the solvent, from the polymer held at constant stress and ΔV^{PTX} is the corresponding volume change. A similar equation applies for constant strain Y.

$$dP/dT = \Delta H^{PTY}/(T \Delta V^{PTY}) \tag{81}$$

Equations 79, 80, and 81 apply to the equilibrium between water vapor and textile fibers as well as between solvent vapor and bulk polymers swollen with solvent.

The remainder of this section is based on: (a) the description of the swelling properties of rubber as described by Treolar (1958, Chapter VII), and (b) the standard requirement for equilibrium, that the free energy at equilibrium be a minimum. The particular free energy involved as the criterion of the minimum depends on the system.

If swollen rubber is treated as a single phase system of unvarying composition, the thermodynamics of its elasticity is the same as for unswollen rubber. A different treatment is required if the system consists of two phases, one component of which may migrate between the phases. The two phases can be a swollen rubber immersed in a swelling solvent or a swollen rubber in equilibrium with the vapor of a swelling agent.

Consider first a rubber body immersed in a solvent, under tension but held at constant extension Y. The system is held at constant temperature and pressure. Phase α is the solvent and phase β is the swollen rubber. The condition for equilibrium is that the sum of the available work from both phases be at a minimum — i.e., that the sum of their free energies be at a minimum. From equation (7) Chapter VI this is,

$$dG^{PT} = dG_\alpha^{PT} + dG_\beta^{PTY}$$
$$= dE_\alpha^{PT} + dE_\beta^{PTY} - T(dS_\alpha^{PT} + dS_\beta^{PTY}) + P(dV_\alpha^{PT} + dV_\beta^{PTY}) \quad (82)$$

However if the swollen body is held at constant tension X instead of at constant extension the model is different and hence the criterion for equilibrium is different. The expression for the available work is then,

$$d(G_\alpha^{PT} + G_\beta^{PTX}) = dE_\alpha^{PT} + dE_\beta^{PTX} - T(dS_\alpha^{PT} + dS_\beta^{PTX})$$
$$+ P(dV_\alpha^{PT} + dV_\beta^{PTX}) + X\,dY_\beta^{PTX} \quad (83)$$

The process by which equilibrium is attained in these systems is the transfer of the solvent into the swollen rubber phase or its reverse. If dm_1 is the amount of solvent transferred from the solvent phase into the rubber phase we can take the partial derivative of equation (82) with respect to m_1. This derivative must be zero at equilibrium. We then obtain,

$$(\partial E_\alpha/\partial m_1)_{PT} + (\partial E_\beta/\partial m_1)_{PTY} - T((\partial S_\alpha/\partial m_1)_{PT} + (\partial S_\beta/\partial m_1)_{PTY})$$
$$+ P((\partial V_\alpha/\partial m_1)_{PT} + (\partial V_\alpha/\partial m_1)_{PTY}) = 0 \quad (84)$$

From equation (83) we can write a similar equation for equilibrium at constant P, T, and X,

$$(\partial E_\alpha/\partial m_1)_{PT} + (\partial E_\beta/\partial m_1)_{PTX} - T((\partial S_\alpha/\partial m_1)_{PT} + [\partial S_\beta/\partial m_1)_{PTX})$$
$$+ P((\partial V_\alpha/\partial m_1)_{PT} + (\partial V_\beta/\partial m_1)_{PTX}) + X(\partial Y_\beta/\partial m_1)_{PTX} = 0 \quad (85)$$

Equations (82) through (85) can be rewritten in a slightly different form by combining various terms. Thus (82) can be converted to,

$$dG^{PTY} = d(E_\alpha + E_\beta)^{PTY} - T\,d(S_\alpha + S_\beta)^{PTY} + P\,d(V_\alpha + V_\beta)^{PTY} \quad (86)$$

This gives a slightly different viewpoint of the meaning of these equations.

Now, consider a rubberlike body which is stretched to an amount Y, held at constant temperature and pressure, and swollen with a solvent in which it is immersed. This body is at equilibrium. This body is then stretched a further amount dY. There will then be an adjustment in the amount of solvent absorbed as the system returns to an equilibrium. This adjustment will lower the free energy slightly. In rubber like bodies the elastic free energy is primarily entropic so one would expect the adjustment after the extension dY to also be primarily entropic. Hence, any absorption of solvent (or likewise any desorption of solvent) caused by the increase in extension dY should be accompanied by an exchange of heat. It should be noted that this is a reversible path as the process is reversed if dY is reversed.

The concepts of the relationships between type of elasticity and absorption or desorption of heat have not been checked against experimental data, however, the above is an example of how detailed thermodynamic reasoning can be used to predict behavior.

An Example from Metallurgy

It is interesting to note that there is a close analog with metals to the swelling behavior of rubber under stress. The permeability and the solubility of hydrogen in palladium and some of its allows is very high. Wriedt (1970) used an alloy of 75% palladium and 25% silver to study the change in the amount of hydrogen absorbed under varying tension and compression. The amount of hydrogen absorbed increased linearly under tension and decreased linearly under compression. A graph of Wriedt's results is shown in Figure 12. The authors report a change in volume of the alloy with the absorption of hydrogen; though, they did not measure that change in volume with the change in stress.

Their data fits the equation,

$$\ln (c_i/c_i^o)_{f_i} = \sigma \bar{V}_i/3RT \qquad (87)$$

where c_i and c_i^o are the concentrations of the hydrogen in the stressed and unstressed metal both at the same fugacity f_i, σ is the uniaxial stress and $\bar{V}_i$ is the partial molal volume of the hydrogen in the metal. This equation fits the behavior of this metal and is not necessarily valid for other types of materials.

Another Thermodynamic Machine for the Production of Work

It is interesting that a machine has been patented (Katchalsky et. al. see references) that uses the thermodynamics of swelling for the production of work. A belt made of a material which swells in a water solution (i.e., lengthens) and shrinks in a salt solution is selected. Collagen is an example of a suitable material for a belt. The belt is passed over two pulleys in the salt solution — one larger than the other. When the belt contracts the torque on the larger pulley is stronger causing the belt to move. This brings a fresh uncontracted section of the belt into the salt bath and moves the contracted section into a bath of pure water where it relaxes. The result is a con-

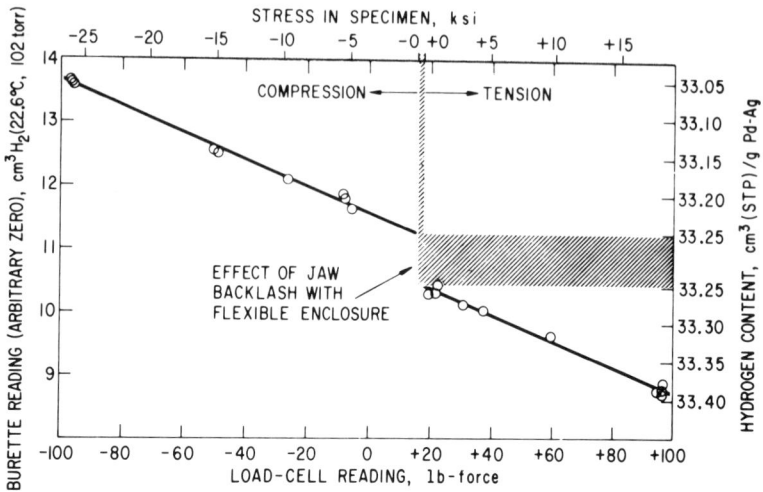

FIGURE 12. The effect of elastic stress upon equilibrium hydrogen content of an alloy of palladium and silver at 75°C and 102 torr of hydrogen gas, from Wriedt (1970).

tinuous motion in one direction. This device can also be thought of as using a concentration gradient to produce work. See Chapter X.

A similar machine has been made which uses a rubber belt. The belt contracts on heating and expands when cooled. Work is produced by a mechanism similar to the above. This is a heat engine just as steam engines and turbines are.

Non-Uniform Stress

A simple non-uniform stress is generated when a long rod is hung from one end. The stress (tension) varies from a maximum at the top to zero at the bottom. The gradient in stress is relatively small unless the body has a very high density.

A second example is a rubber rod fastened to the bottom of a container and then immersed in mercury. There are two gradients — one in pressure with a maximum at the bottom and atmospheric at the top — a second in tension which is also a maximum at the bottom and zero at the top. These stress gradients are small compared to what can occur under bending, twisting and other actions.

Consider the first example of a rod hung at its top but the material in the rod has a very high density so the stress gradient is relatively large. Take a plane across that rod. This is a body A at pressure P, temperature T, stress X and stress gradient dX/dx. Are the properties of body A the same as that of a body B which is at the same pressure, temperature and stress but in which there is no stress gradient? By definition they are not and dX/dx is one of the independent variables of state. The equation of state is,

THE THERMODYNAMICS OF ELASTIC DEFORMATION

$$B = f(P, T, X, dX/dx) \tag{88}$$

We do not have measurements of the effect of the stress gradient on the dependent variables of state; however, be they large or very small it is self evident that the gradient is an independent variable of state.

The equation of state in (88) applies to a plane in the body. It could also apply to a line, or if twisting and bending stresses exist, to only a point in a body.

Since any body with a gradient is not uniform, its extensive properties — such as, volume, internal energy, entropy, etc. must be calculated by the summation of the properties of their infinitesmal components (planes, lines, or points). Hence, the thermodynamic functions are also summations of those of the bodies infinitesmal components. Although the conditions for equilibrium can be given in principle, the actual statement of these conditions for equilbrium in the presence of gradients may become complex. Gradients will be discussed further in Chapter X.

Chapter VIII
The Separability of the Reversible and Irreversible in Thermodynamic Processes

Thermodynamic or Thermostatics?

Classical thermodynamics is sometimes referred to as "thermostatics". In one sense this is correct as all bodies are described in terms of their being at equilibrium. However, labeling thermodynamics "thermostatics" because the concept of "equilibrium" underlies the equations used is bad terminology; for, it implies that its subject matter is completely static. Classical thermodynamics is concerned with processes which move. It had its origin in machines which convert heat to work (steam engines). Internal combustion engines and turbines have replaced the old steam engines; but, their operations are still described in terms of classical thermodynamics. Since its origin there are many other dynamic processes which have been added to the province of classical thermodynamics — batteries, refrigerators, the blast furnace, the chemical processes, etc. This book is about the thermodynamics of viscoelastic deformation — a dynamic process. Thermodynamics is the preferred name on both an historical and a conceptual basis.

A Non-Static Thermodynamics

It is desirable to take a viewpoint on thermodynamic processes which emphasizes the dynamic rather than the static approach. A key to this viewpoint is that the reversible and irreversible are separable in all real thermodynamic paths. In other words all thermodynamic changes can be divided into their reversible and irreversible components. This viewpoint (or emphasis) makes only minor differences in the subjects traditionally treated in (two independent variables of state) classical thermodynamics. However, when the concept is extended into the areas traditionally considered as "irreversible thermodynamics" and where the system has more than two independent variables of state, the significance of the concept becomes apparent.

The concept of the separability of the reversible and irreversible components of processes is old. Thomson (Lord Kelvin) used it in his treatment of the thermoelectric effects. His results were correct though the method is generally considered suspect and not generally applicable. When the systems are more carefully described and analyzed, the validity and utility of the principle of the separation of the reversible and irreversible processes becomes apparent.

This separability does not mean that the two types of processes are not coupled together in some way, for the irreversible frequently influences the reversible and vice-versa.

The Partitioning of Work in the Spring and Dashpot Models

In the spring and dashpot models any work must be done against either the springs (reversible) or against the dashpots (irreversible). Hence, the spring and dashpot models characteristically partition all work into reversible and irreversible. In order to evaluate this property of these models their behavior will be examined more closely.

Consider the simplest model — one spring and one dashpot in parallel (Figure VII-4a). These need not be linear. Figure 1 shows a force displacement diagram for this model as it is first extended and then allowed to recover, doing work against a restraining force. The dotted lines represent the external force acting on the system. The solid line represents the force exerted by the spring which is the same whether the displacement is increasing or decreasing. The external forces are greater than the spring force when the displacement is increasing, and less than the spring force when the displacement is decreasing. The area enclosed by the dotted line is equal to the irreversible work done during one cycle. The area under the solid line represents the reversible work done on the spring; however, the reversible work is only partially recovered because of the action of the dashpot during relaxation.

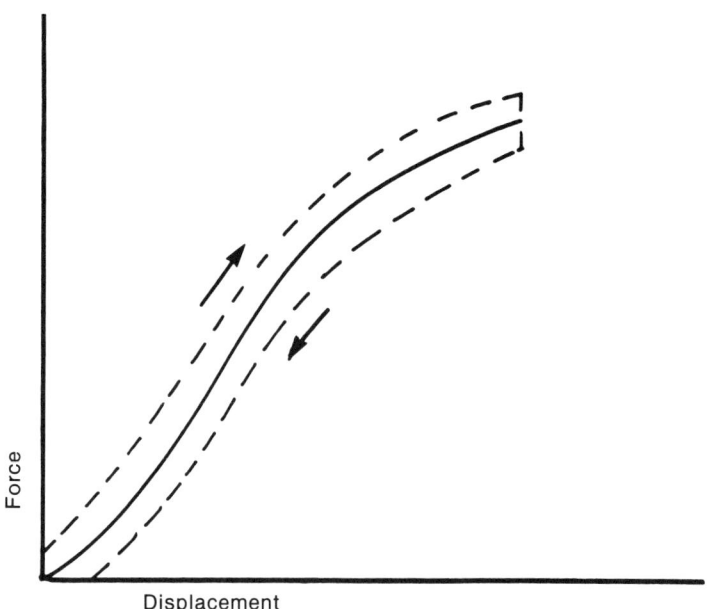

FIGURE 1. Force displacement diagram for a single spring in parallel with a dashpot. The solid line represents the force at equilibrium which is also equal to the force on the spring alone. The dotted lines represent the force during a moving cycle. The spring and dashpot are not linear.

The single spring and dashpot model is simple. When movement stops the external force immediately changes so that it becomes equal to the force of the spring (since the force exerted by the dashpot is then zero). More complex models exhibit different behavior. Consider two pairs of parallel springs and dashpots connected in series as in Figure VII-5. Consider that in the first spring and dashpot pair, the spring is weak and the dashpot is strong. The second pair has the reverse relationship — a strong spring and a weak dashpot. If this unit is displaced rapidly and then held at a constant displacement the strong spring will absorb the work of the initial displacement; then, while the displacement is held constant, the strong spring will do work against the weak spring and the two dashpots, gradually decreasing the externally exerted force. This decay of the external force and the internal reversible stored energy will continue until a three variable equilibrium state is attained. See Figure 2. This conversion of a part of the stored elastic energy to a non-available state by an internal process will be called internal relaxation.

The concepts of internal relaxation can be applied to a force displacement diagram (analogous to Figure 1). Such a diagram is shown in Figure 3. The solid line now represents the three variable equilibrium state when all internal relaxation processes have been completed for each position on the line. The work done against this model, when the displacement is increasing, is the area under the upper dotted line in Figure 3. This work can be partitioned into three parts:

(a) The irreversible work done directly against the internal resistance of the dashpots (the area between the upper dotted line and the upper dashed line).

(b) The work temporarily stored in the springs as reversible work, but which can later be converted to irreversible work by the process of internal relaxation (the area between the upper dashed line and the solid line).

(c) The work stored in the springs and not releasable by the internal relaxation process (the area under the solid line).

When the displacement is allowed to decrease and do work against a restraining force the stored energy is released and partitioned into three parts analogous to the above:

(d) The work done directly against the external restraining force (the area below the lower dotted line).

(e) The work done directly against the internal resistance of the dashpots (the area between the dashed line and the dotted line).

(f) The energy not released immediately but released later by internal relaxation (the area between the lower dashed line and the solid line).

For any one model the solid line in Figure 3 remains in a fixed position; however, the positions of the dotted and dashed lines depend on the rates of displacement and can vary from cycle to cycle. Figures 1 and 3 show the partitioning of work (or energy) into reversible and irreversible on the spring and dashpot models.

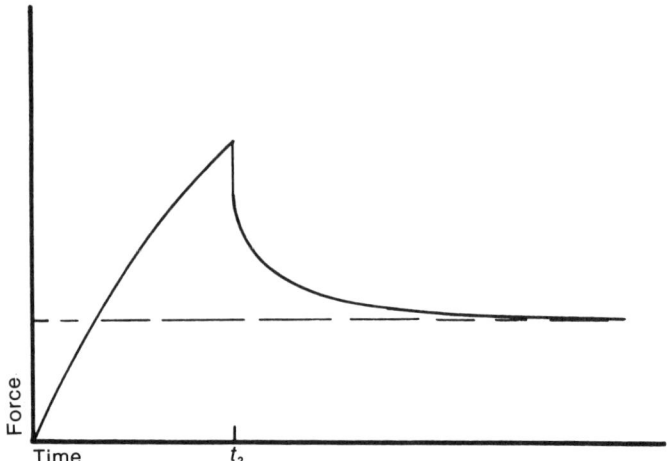

FIGURE 2. Diagram of focus versus time with two pairs of parallel springs and dashpots connected in series as per Figure VII-5. The displacement is first increased and then stopped and held at time t_2. The force increases with the increasing displacement but when displacement is held constant the force decays asymtotically approaching the three variable equilibrium condition.

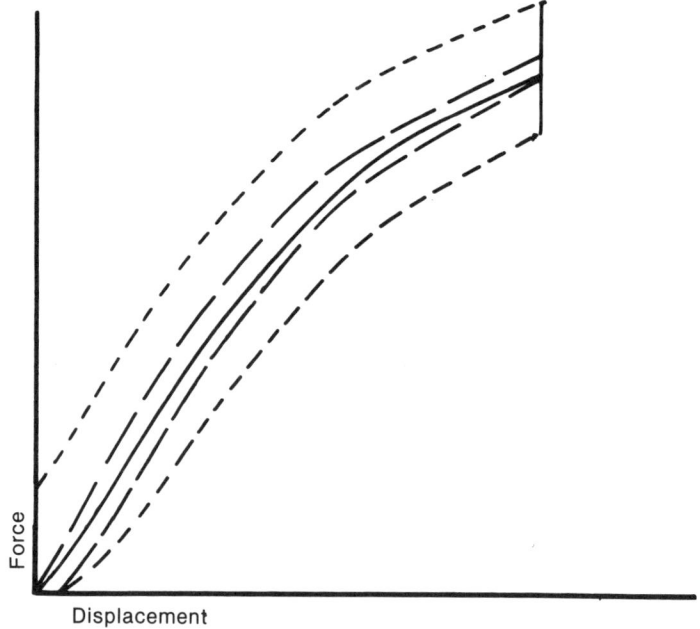

FIGURE 3. Force displacement diagram for a series of two or more units of springs and dashpots in parallel as in Figure VII-5. In such a set of springs and dashpots an internal relaxation may occur after its displacement has become fixed. Such an internal relaxation is diagramed in Figure 2. This figure differs from Figure 1 in that it diagrams an area (surrounded by dashed lines) which represents the energy which may be made irreversible by internal relaxation. As in Figure 1 the solid line represents the three independent variable equilibrium line.

Molecules as Springs and Dashpots and Internal Work

One model for a viscoelastic solid is a large number of molecularly sized springs and dashpots, all linked together. However, this model is inadequate. It is improved by assuming that this model contains two kinds of springs — a kinetic energy spring and a potential energy spring — linked parallel to each other and parallel to a molecular dashpot. See Figure VII-10. The kinetic energy spring is a special kind of spring; the *force it exerts at any one extension is proportional to the absolute temperature*. On the other hand the potential energy spring is an ordinary spring whose force is a function of its extension and not a function of any other variable. If the temperature is held constant the effects of temperature on the kinetic energy spring are not brought into play, and two ordinary type springs in the model of Figure VII-10 are adequate. The kinetic energy spring could also be called the entropy spring if conditions were limited to constant temperature.

The kinetic energy spring is an analog of the exertion of force by thermal motion. With a gas the thermal motion of the molecules exerts a pressure by which work is exchanged with its surroundings. With polymers the thermal motion of the polymer chains exert a force to permit the polymer chains to attain the average configuration of the unstrained body. Work done in imposing a strain on the kinetic energy spring is work done against the thermal motion of the polymer chains. If the temperature remains constant during the imposition of the strain, heat is removed, giving an actual change in entropy. If the conditions are adiabatic during the imposition of the strain the kinetic energy spring is displaced releasing heat and raising the temperature of the body; but, there is no change in the entropy of the body. The kinetic energy spring is reversible so that all work done on it is converted to heat, and all work done by it uses heat.

The potential energy spring is an analog of the action of the intermolecular forces. The mechanical spring is a more perfect analog of the intermolecular forces (than of the forces caused by thermal motion) for the intermolecular forces are a function of displacement only. Work done against this spring is stored as intermolecular potential energy. The force exerted is known as internal pressure in gases and internal tension in rubberlike bodies under tension. See equations IV-19 and VII-13, 18.

Properties of the Molecular Kinetic Energy Spring

The kinetic energy spring has special properties which no real metal coil spring has. These properties are basic to thermodynamics for they relate work to heat and temperature — the area which distinguishes thermodynamics from ordinary mechanics. These properties are,

(a) The force exerted by the spring is proportional to the absolute temperature as well as being a function of displacement (or strain). We can then write,

$$X = X_s - X_i \quad (1)$$

$$X_s = X + X_i = T f(x) \quad (2)$$

where X is the total tension in a rubber body, X_s is the kinetic component of that tension, X_i is the internal tension, and x is the displacement of the kinetic energy spring.

(b) The kinetic energy spring has the special property that all work done on it is converted to heat and all work done by it uses heat. If the temperature is held constant, heat must be exchanged with the surroundings and there is a change in entropy. If the conditions are adiabatic the heat remains in the body. There is then no change in entropy even though the kinetic energy spring is extended by the imposition of the strain.

(c) It has the property such that certain cycles can convert heat to work or act in their reverse as refrigeration cycles. The Carnot cycle is an example of such a cycle. This cycle will be discussed in detail later in this chapter.

(d) The kinetic energy spring has the special property that it can sometimes relax or partially relax without changing its extension — i.e., it no longer exerts a force although it is still extended. For a rubberlike body this results from local rearrangements of the average positions of the polymer chains, by means of which the molecular mechanism for the conversion of heat to work is lost. This relaxation converts the heat which has been generated from the extension of the kinetic energy spring from reversible to irreversible. Partial relaxations of this type can occur with rubberlike solids for they are restrained from going to complete relaxation (i.e., an exertion of zero force) because of the restraining cross-links.

The Validity of the Spring and Dashpot Model

The spring and dashpot models are valid if the kinetic energy spring has the special properties listed above and the springs are both nonlinear. If the kinetic energy spring is considered to be a nonlinear steel coiled spring the model has validity under limited conditions.

There are two different specific conditions for which the steel spring model is valid. The first condition is isothermal. In an isothermal cycle there is no net work generated by the conversion of heat energy into mechanical work for temperature differences (such as in the Carnot cycle) are required for such a conversion. The statement that a completely isothermal cycle generates no net work is known as Moutier's theorem (Guggenheim 1967, p. 45). The description of the cycle for a spring and dashpot model is then equally valid for isothermal cycles on real rubber bodies.

The second specific condition for which no net work is generated in a cycle is adiabatic — for then no heat is transferred from which work can be obtained. However, it happens that a truly adiabatic cycle cannot be obtained with a viscous element in a rubber body. Heat is generated by the irreversible viscous action which prevents the return of the body to the original state by an adiabatic path. This problem can be solved by the use of a pseudo-adiabatic cycle in which all irreversible heat is removed at the time and temperature of its generation. Such a cycle would be very difficult to generate in practice but it is a very simple thought experiment. The description of the cycle in Figures 1 and 3 for springs and dashpots is then equally valid for pseudo-adiabatic cycles on real rubber bodies.

The Interactions of the Three Elements in the Model

The two springs and the dashpot interact with each other in various ways, and these interactions should be closely examined.

In most real rubbers the potential energy spring aids the external forces in the straining of the kinetic energy spring. In consequence the heat generated in placing a strain on the kinetic energy spring, is equal to the sum of the external work plus the internal work of the potential energy spring.

If the two springs are allowed to relax doing their maximum external work, (i.e., no internal relaxation) their actions are completely reversible; and the heat is converted to internal work against the potential energy spring and to external work against outside forces. If the two springs are allowed to relax without doing any external work, the kinetic energy spring (i.e., the kinetic motion of the molecules or polymer chains) will force the polymer chains into a more random position at which the intermolecular potential energy is higher. This uses heat, thus cooling the body. Equivalent effects take place if the relaxation is by means of local rearrangement of the polymer chains with extension remaining constant or by means of a relaxation in which the body recovers its original shape without doing external work.

Any action of the potential energy spring should be regarded as completely reversible. If it opposes the kinetic energy spring as it does in rubberlike bodies and real gases it always exchanges energy with the kinetic energy spring. If it does not oppose the molecular force spring as with non-rubberlike solids it must work against external and frictional (internal dashpot) forces.

Any work done on the dashpot is purely irreversible. Although the heats generated in doing work on both the dashpot and the kinetic energy spring are both equal to the work done on them, the difference is that the kinetic energy spring provides a mechanism for reversability.

If the kinetic energy spring does work on the dashpot the heat used up is equal to the heat generated so there is no change in temperature. The effect is that the heat loses its reversibility. If the kinetic energy spring relaxes without doing any work the same amount of heat is made irreversible. The two processes are equivalent.

If the potential energy spring does work on either the dashpot or the kinetic energy spring the effect is that an equivalent amount of heat is generated. The temperature rises if heat is not removed.

A Carnot Cycle with Rubber

The classic Carnot cycle uses a gas as the body which is cycled. During the expansion there is a constant temperature path and then an adiabatic path. During the compression portion of the cycle there is also a constant temperature path and then an adiabatic path. Figure 4 shows the cycle in terms of three different sets of coordinates: (a) temperature and volume, (b) temperature and entropy, and (c) pressure and volume.

A Carnot cycle can also be constructed using rubber as the body which is cycled. Figure 5, (a), (b) and (c) shows the three charts corresponding to those in Figure 1. The differences are that pressure is replaced by force X and volume by displacement Y. The two variables X and Y may refer to any elastic displacement such as that in shear, tension or compression. The main difference between the gas and solid cycles is that a gas increases its volume when allowed to do work against its surroundings while an elastically displaced solid body decreases its displacement when allowed to do work against its surroundings. Hence, in Figure 4 the sequence of points around the cycle for (a), (b) and (c) are all in the clockwise direction: while in Figure 5 the sequence of points for (a) and (c) are in the counter clockwise direction though (b) is still in the clockwise direction. When the Carnot cycle is used for refrigeration the sequence of all the points in both figures is reversed.

In Figures 4 and 5 the solid lines show the limits of real paths as the irreversible components go to zero. That is, no friction, no irreversible flow of heat to a lower temperature, and no internal relaxation has occurred. Under these conditions the net work done (w) is the area enclosed by the solid lines in the (c) graphs in both figures. The net work is also equal to the difference between the heat absorbed at the high temperature and the heat given up at the low temperature, or

$$w = q_h + q_c = T_h \Delta S + T_c \Delta S \qquad (3)$$

hence, the area enclosed by the solid lines in the (b) figures is also equal to the work done.

The change in entropy for the system is zero, i.e.

$$q_h/T_h + q_c/T_c = 0 \qquad (4)$$

and the efficiency of the conversion of heat to work is,

$$(T_h - T_c)/T_h = \text{efficiency.} \qquad (5)$$

The above cycles for both elastomers and gases are ideal cycles as they have no irreversible components.

The Carnot cycle using rubber as the cycling body has been introduced in part because it is a continuation of Chapter VII on the thermodynamics of elastic deformation. However, the primary reason for its introduction here is that solids are *visco*elastic, and hence demonstrate the effect of irreversible work on one of the classic examples of thermodynamics. The viscous component of the compression and expansion of gases is very small if it exists at all.

When a Carnot cycle is attempted with a *visco*elastic body the net work obtained is less than the ideal. A plot of the external force during such a cycle is shown by the dotted lines in Figure 5c. The dotted lines for the adiabatic expansion and adiabatic compression are analogous to those in

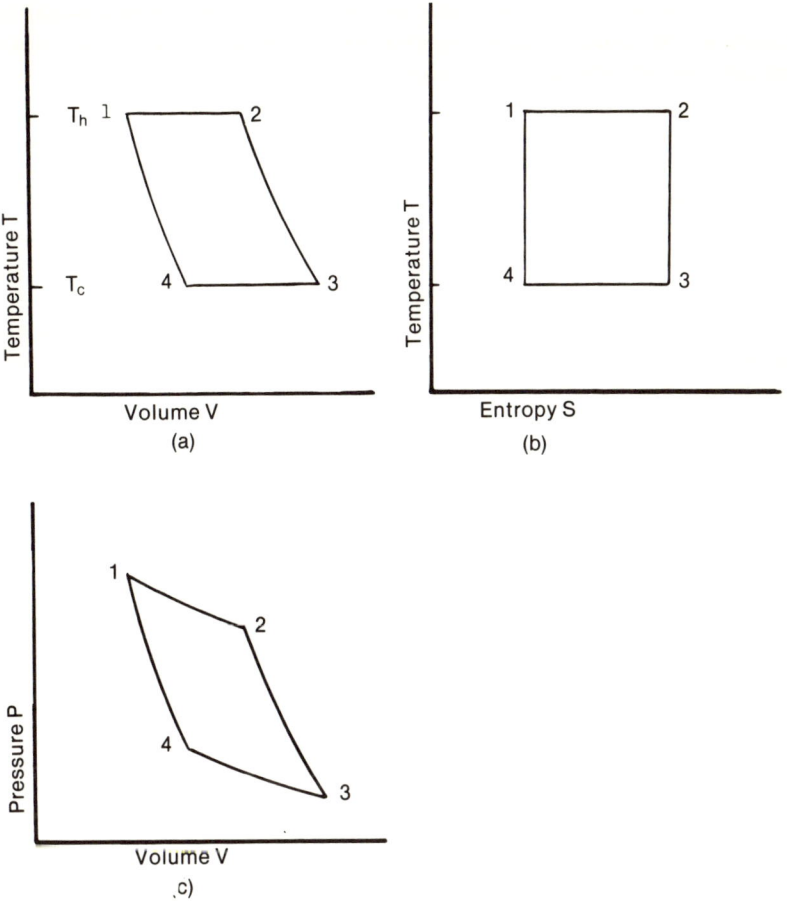

FIGURE 4. Diagrams of the Carnot cycle using gas as the cycling body. (a), (b), and (c) show the same cycle with three different sets of coordinates. The work done in one cycle is equal to the areas enclosed in both (b) and (c).

Figures 1 and 3 if they are presumed to be pseudo-adiabatic in which the irreversible heat generated by viscous flow and irreversible relaxation is removed at the time and temperature of its generation.

The solid lines in Figure 5 (c) can be viewed in three different ways,
   (a) They are the paths taken when the displacement is so slow that friction and molecular relaxation are negligible.
   (b) They are the limits of the paths as the irreversible processes approach zero.
   (c) They represent hypothetical reversible paths between two equilibrium points which can be used for thermodynamic calculations no matter what the actual paths are.

These three viewpoints are equivalent in one sense, yet much more is implied in (c) in that they provide paths for thermodynamic calculations which are independent of the actual path.

All paths between two points are not completely thermodynamically equivalent. In Figure 5 point 3 can be reached in two different ways on the paths shown. First it can be reached by going through (2), and second it can be reached by reversing the direction and going through point (4). There is a big difference in the external work done and in the heat flows on the two paths; however, the properties of the body (in equilibrium at point 3) are the same no matter what path is taken. The thermodynamic changes in a body when it goes from one equilibrium state to another can be calculated by using any hypothetical reversible path between those states. The thermodynamic changes in the body will be the same no matter what the path chosen, even though heat flow and external work vary widely with the paths chosen.

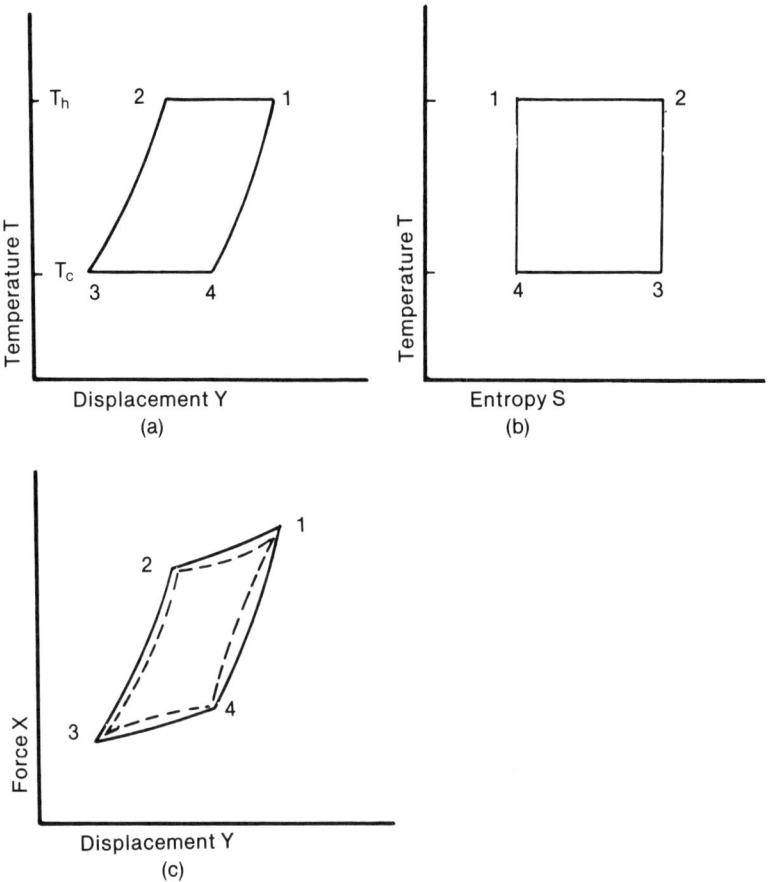

FIGURE 5. The solid rubber lines show the Carnot cycle using a non-viscous rubber as the cycling body or the reversible component of the force using a viscous rubber. The dotted lines in (c) are a plot of the external forces when using a viscous rubber. The area between the dotted lines and the solid lines is the irreversible work.

The Reversible and Irreversible Components of Gas Expansion

The process of mixing by dissolving one material in another is similar to the expansion of a gas into a vacuum. The thermodynamics of the two processes can be developed in similar ways.

Consider the expansion of a gas against an opposing pressure while held at constant temperature with a heat source. This is a completely reversible path. From the first law we can write for an incremental amount of heat absorbed by the gas,

$$\delta Q = T\,dS = dE + P\,dV \qquad (6)$$

This can be rewritten more explicitly as,

$$T(\partial S/\partial V)_T = (\partial E/\partial V)_T + P \qquad (7)$$

or

$$(\partial S/\partial V)_T\,dV = (1/T)(\partial E/\partial V)_T\,dV + (P/T)\,dV \qquad (8)$$

Now consider an irreversible path between the same two points in which the gas expands into a vacuum (instead of against an opposing pressure) while held at constant temperature. A close examination of the processes involved shows that the amount of heat absorbed is equal to $(\partial E/\partial V)_T dV$ in equation (7). It is less than the amount of heat absorbed over the reversible path by the amount $P\,dV$.

It is better to compare the above two reversible and irreversible paths in terms of the entropy changes as given by equation (8). Entropy changes are divided into two parts. In the completely reversible path both parts of the entropy change are reversible. In the second path the first term remains reversible but the second represents an irreversible entropy change. Hence, the difference is entropy at the two points can be divided into two parts, one of which is always reversible and one of which can be either reversible or irreversible depending on the path. We can then write,

$$dS = dS_{\text{rev}} + dS_{(\text{rev or irr})} \qquad (9)$$

The entropy change of the gas is the same no matter what the path between the initial and final states, for the entropy is one of the dependent variables (i.e., a B) in the equations of state. As such it is a property of state which is determined by those initial and final states and not the path between them. Equation (9) is also valid for solid bodies.

The integral of equation (7) is the equivalent of the equation for internal work, (equation VII-21) for it can be converted to that from the Maxwell relations,

$$(\partial S/\partial V)_T = (\partial P/\partial T)_V \qquad \text{VII-54}$$

The Reversible and Irreversible Components of Mixing

The mixing of two solvents is similar to the expansion of a gas in that the entropy changes involved can be divided into two parts one of which is always reversible and the other of which may be either reversible or irreversible. Consider two different liquids (1), and (2), which are mutually soluble in each other. If these two liquids form an ideal solution, their mixing will cause no change in total volume, and there will be no evolution or absorption of heat. There will only be an irreversible increase in entropy. If, however, the mixing process is carried out with the use of a membrane which is permeable to liquid (1) but not to liquid (2), the mixing process can be made reversible. When the membrane is placed between the two liquids an osmotic pressure Π is formed which can perform reversible work in the amount of

$$\delta Q = T\, dS = \Pi\, dV_2 \qquad (10)$$

where dV_2 is the change in volume of volume (2) as liquid (1) passes through the membrane.

If the two liquids do not form an ideal solution, heat may be generated or absorbed and there may be a change in total volume when the two liquids are mixed. These effects are all additive to the first so that the heat absorbed becomes,

$$\delta Q = T\, dS = \Pi\, dV_2 + (\partial(E_1 + E_2)/\partial V_2)_T\, dV_2 + P\, d(V_1 + V_2) \qquad (11)$$

or

$$dS = (\Pi/T)\, dV_2 + (1/T)(\partial(E_1 + E_2)/\partial V_2)_T\, dV_2 + (P/T)\, d(V_1 + V_2) \qquad (12)$$

where the entropy change on mixing has been partitioned into a component which is always reversible and a component which may or may not be reversible as in equation (9).

If the two liquids become cooler on simple mixing (i.e., without the semipermeable membrane) the analogy to the expansion of gases is complete for then the internal energy must increase on mixing, if the temperature is kept constant. If heat is released on simple mixing this indicates that the internal energy decreases on mixing (at constant temperature), and hence, that the molecules of the two materials must form some type of association with each other on mixing. Such an association could be the formation of a chemical compound but it is likely to be a looser association. This assumes of course that effects from any volume changes are negligible.

The Analogy Between Free Energy and Energy Stored in Springs

Free energy has been defined here (in Chapter V) as the maximum available work as a body goes from a state (1) to a lower free energy state (2). It has been pointed out that the free energy available depends upon the restraints on the system. See Chapter VII equations (3) through (6). The free energy is also the minimum work necessary to go from a state (2) to a state (1) of a higher free energy within the same stated restraints. Table I lists some of the various energies with their associated restraints.

The work available from the springs in the spring dashpot models represents a free energy; however, there is no simple way of showing the differences in the various thermodynamic free energies shown in Table I. The spring model can be considered to represent any of these free energies.

Table I. Some of the free energies with their restraints.

Restraints which are held constant	The associated free energy
Two restraints	
T,P	ΔG^{PT}
V,T	ΔA^{VT}
V,S (adiabatic)	ΔE^{VS}
P,S, (adiabatic)	ΔH^{PS}
Three restraints (partial list)	
$T,P,X,$	ΔGG^{PTX}
T,P,Y	ΔG^{PTY}
P,S,X (adiabatic)	ΔHH^{SPX}
P,S,Y (adiabatic)	ΔH^{SPY}

The spring and dashpot model with the two springs is particularly suitable for the representation of the elastic properties of bodies and the accompanying free energy changes. The springs are also suitable for the representation of other types of free energies such as that involved in mixing, etc. Most of these other types will not include a component analogous to the dashpot; however, the concept of two springs which may oppose each other is still useful, for any work done must be derived from the intermolecular potential energy and/or from the action of the thermal motions of the molecules.

It is also possible for a body to have a free energy with the restraints of constant energy E, constant volume V, and constant extensive variable of state Y. This free energy can be dissipated by irreversible processes. If the free energy is to be converted to work, this work must remain within the system and be considered a part of the systems energy E. The energy in the system can then be divided into three parts — intermolecular potential energy E_p, the molecular kinetic energy E_k, and a work of unspecified type E_w. For visualization purposes E_w can be considered as work done against a gravitational, or electrical, or magnetic field all of course within the isolated system. The maximum E_w obtainable from a closed system is then a free energy. Three examples of free energy in isolated systems (i.e., constant E, V and Y) follow.

1. Composition differences. With liquids the maximum available work is the integral of the osmotic pressure times the associated differential volume,

$$E_w = \int \Pi \, dV_2 \qquad (13)$$

With gases differences in composition are associated with differences in partial pressure. If partial pressure is substituted for osmotic pressure equation (13) holds. With solids molecular mobility is so low that the measurement of any pressure analogous to the osmotic pressure is probably not possible; however, the maximum available work can still be stated in terms of a partial pressure analogous to the osmotic pressure. Any actual work done by osmotic-like pressures in gases, liquids and solids will be done by the thermal motion of the molecules and hence lower the temperature.

2. Differences in Temperature. The maximum efficiency obtainable from an engine which transfers heat from a high temperature source to a lower temperature acceptor is,

$$(T_h - T_c)/T_h \tag{14}$$

If we consider the simple case of two bodies at different temperatures, heat can be taken from one and partially converted to work. The rest of the heat will be discharged to the body at the lower temperature. This process lowers the higher temperature, and raises the lower temperature. The maximum work is then,

$$E_w = \int ((T_h - T_c)/T_h) \, dq \tag{15}$$

where both T_h and T_c are variable temperatures. The integration is carried out over the process which brings the system to a uniform temperature.

3. Chemical Reactions. It is presumed that chemical reactions are caused by the attractions of certain unlike molecules to each other. Under this simplified assumption chemical reactions then result in a lowering of intermolecular potential energies. If this loss in intermolecular potential energy is reversibly converted to work E_w, there may also be changes in the kinetic energy of the molecules associated with the changes in the degrees of freedom of motion these molecules. We can then write,

$$E_w = \Delta E_p^{VY} + \Delta E_k^{VY} \tag{16}$$

Equation (16) is also valid for examples (1) and (2) above; however, in those cases the major energy changes are in E_k rather than in E_p.

The Non-Additivity of the Free Energies

One may ask if the various types of different free energies of Table I are additive. This may be investigated as follows.

Consider a sample or rubber which is ideal in that there are no viscous components when it is deformed. Its volume is V_1. If this is stretched at constant pressure P_1 its free energy becomes

$$\Delta G^{P_1 T} = \Delta E^{P_1 T} - T\Delta S^{P_1 T} + P_1 \Delta V^{P_1 T} \tag{17}$$

It is further stretched but now held at a constant volume V_2 the supposed additional free energy is,

$$\Delta A^{V_2 T} = \Delta E^{V_1 T} - T\Delta S^{V_2 T} \tag{18}$$

The rubber body then may be allowed to return to its original condition over different reversible paths by first keeping the pressure constant at P_2 until volume V_1 is reached when the recovered work is,

$$\Delta G^{P_2 T} = \Delta E^{P_2 T} - T\Delta S^{P_2 T} + P_2 \Delta V^{P_2 T} \tag{19}$$

Then the rubber is allowed to recover further doing external work at constant V_1 until the initial state of P_1, V_1 is reached. This work is,

$$\Delta A^{V_1 T} = \Delta E^{V_1 T} - T\Delta S^{V_1 T} \tag{20}$$

If the different types of free energies are additive the reversible work between P_1, V_1 and P_2, V_2 is the same regardless of the path; i.e., the sum of the equations (17) and (18) must equal the sum of the equations (19) and (20). Since E, and S are properties of state the differences in these terms are equal irrespective of the path between the states (1) and (2). V is also a property of state so that ΔV is the same for both halves of the cycle. The difference between the input and output paths is then

$$(P_2 - P_1)\Delta V \tag{21}$$

The work done on and received from the stretching and recovery of the rubber is given by (21). This is a cycle in which the stretching and recovery of the rubber pumps energy into the higher pressure level or when reversed receives energy from the higher pressure level. If the cycle had included two temperature levels at which heat had been exchanged with heat sinks and sources, we would then have a cycle which generates work from thermal energy — such as occurs with the Carnot Cycle.

It has been shown that the reversible work between any two thermodynamic points is not always the same, and hence the free energies of Table I are not additive.

The Relationships Between the Free Energies

Equilibrium is attained when the available work (the free energy) is at a minimum. At such a point the derivative of the free energy (within the restraints suitable for that free energy) is equal to zero — for that is the test for a minimum. The partial derivatives of the four free energy functions are given by equation (VII-43),

$$(\partial H/\partial Y)_{PS} = (\partial A/\partial Y)_{TV} = (\partial G/\partial Y)_{PT} = (\partial E/\partial Y)_{SV} = -X \qquad \text{VII-43}$$

All of these derivatives are equal to $-X$ which must then be zero at equilibrium for two independent variables of state. This is what should be expected, when X is stress for an elastic body, or when X is voltage for a galvanic cell. All four measures of free energy give the same answer for equilibrium even though the restraints on the system are different for each.

The concept of chemical potential is useful to show another relationship between the free energies. The equation (V-34) may be rewritten in terms of two independent variable thermodynamics

$$\begin{aligned}\mu_i &= (\partial E/\partial m_i)_{SVm_j} = (\partial A/\partial m_i)_{TVm_j} \\ &= (\partial H/\partial m_i)_{SPm_j} = (\partial G/\partial m_i)_{PTm_j}\end{aligned} \tag{22}$$

If only one component is considered this equation becomes,

$$\mu = (\partial E/\partial m)_{SV} = (\partial A/\partial m)_{TV} = (\partial H/\partial m)_{SP} = (\partial G/\partial m)_{TP} \tag{23}$$

Summary and Discussion

A number of examples in which reversible and irreversible processes are quantitatively separable have been described. Most of the examples involve elastic deformation, but other examples are included to show their thermodynamic analogy to elastic deformation. The separability of the processes is described first in terms of springs and dashpots, and then proceeds through molecular actions to thermodynamic equations. A discussion of free energies has been included. The minimum amount of work done which is necessary for a given elastic deformation is a free energy. It is also equal to the maximum amount of work reversibly obtainable from that elastic deformation. The magnitude and type of the free energy of an elastically deformed body, however, depends on the restraints on that body.

When a body goes from equilibrium state (1) to equilibrium state (2) the differences in the thermodynamic properties of the body between the two states are the same — no matter what path is taken between the two states. However, the entropy change of the system may involve varying amounts of both irreversible components and reversible work, the amounts of which depend on the actual path taken.

This chapter has been concerned with examples taken from "equilibrium" thermodynamics — though the equilibrium is primarily that of three independent variables of state instead of two. In the next chapter these same methods will be applied to "steady state" thermodynamics, where the thermodynamic paths go from steady state (1) to steady state (2). This will involve thermodynamic variables which may be coupled; but, this does not imply that the reversible and the irreversible are not separable in the thermodynamics of the steady state.

Chapter IX
The Thermodynamics of Viscoelastic Fluids

Liquids are usually thought of as viscous bodies; however, they are also elastic, i.e., they are viscoelastic bodies. The elastic properties of a few liquids first came to the attention of rheologists because their elasticity was associated with some highly visible unusual flow properties. Since the time of the first interest in a few special liquids, more sophisticated methods for the study of viscoelasticity have been developed. Interest has spread in particular to polymer solutions and polymer melts where the concepts of viscoelasticity have both applied and theoretical uses.

The elasticity of some liquids is very similar to the elasticity of rubber; and it will be shown that many of the same thermodynamic equations apply to both. However, the thermodynamics of viscoelastic fluids is a thermodynamics of steady states rather than a thermodynamics of "equilibrium" states.

The Maxwell Model for Viscoelastic Fluids

A spring and dashpot in series is a one dimensional model, whose mechanical behavior is an analog to that of viscoelastic fluids. It is called the Maxwell model and is illustrated in Figure 1.

When a constant force X is exerted on the Maxwell model, it assumes a steady state — in which energy is stored in the spring is constant, and energy is dissipated by the dashpot at a constant rate. If the movement is suddenly stopped, the extension of the spring will gradually decrease, and its energy will be dissipated in the dashpot.

If a simple harmonic motion is imposed on the Maxwell model, the response will depend on the frequency of the cycles. A relatively high frequency will result in the displacement of the spring only, i.e., only reversible work is done. A plot of the force versus displacement will be a straight line. A very low frequency will result in the displacement of the dashpot only, i.e., only irreversible work is done. A plot of force versus displacement will be a circle (when the appropriate scaling factors are used). Intermediate frequencies will result in displacements of both the spring and the dashpot; and a plot of force versus displacement will be an ellipse with its major axis at an angle to the displacement axis. The particular frequencies at which the above characteristics show depend on the constants of the spring and the dashpot.

The above discussion assumes: the displacement of the spring is proportional to the force (Hook's law), and the rate of displacement of the

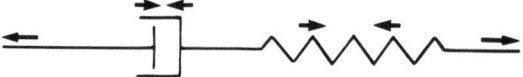

FIGURE 1. The simplest spring and dashpot model which represents the properties of a viscoelastic fluid — known as the Maxwell model.

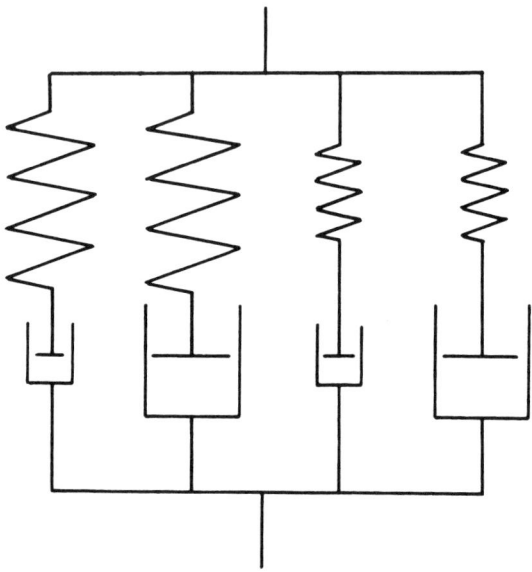

FIGURE 2. More complex responses of linear viscoelastic fluids can be represented by a Maxwell models in parallel — each spring and dashpot with a different constant.

dashpot is proportional to the force (Newtons law of viscous flow). This is the linear model. Such a model can be elaborated to represent more complex responses by the use of a number of Maxwell units in parallel — all with different constants.

The model of linear springs and linear dashpots is particularly suitable for mathematical manipulation and interpretation. Such a model is applicable where displacements are small, and is suitable for use when transducers are used to study polymers and polymer solutions. The variation in response with frequency has a molecular interpretation for these polymers (Ferry, 1970).

The concept of simple springs in arrangements of Maxwell units is not adequate for thermodynamics. It is better to consider all springs replaced by two non-linear springs — one a molecular kinetic energy spring and one an intermolecular potential energy spring as in Figure 3. This is analogous to the model for solids, Figure VII-8. The linear model is not important in thermodynamics, for viscoelastic fluids are not linear when undergoing large deformations.

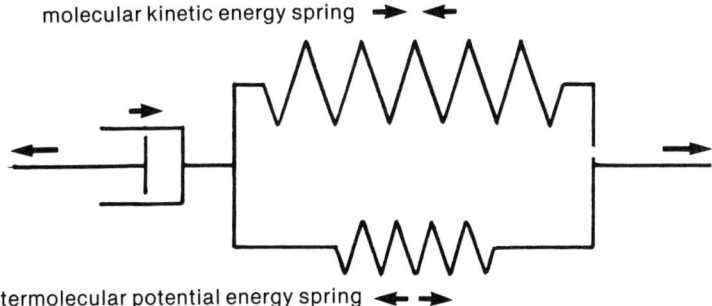

FIGURE 3. When the viscoelastic fluid is deformed at constant temperature the elastic component of the deformation can be represented by two parallel springs — the molecular kinetic energy spring and the intermolecular potential energy spring — connected to a dashpot in series with them.

The spring and dashpot model implies that the reversible and irreversible processes are separable in fluids. The two springs imply there are two types of reversible processes within a viscoelastic fluid. One — the kinetic energy spring — represents the ability of the molecules to do or receive work by means of their kinetic energy, i.e., the heat-work relationship. It also represents the path for the molecular relaxation processes by means of which reversible stored energy is converted to irreversible energy. The second spring — the intermolecular potential energy spring represents the ability of the intermolecular forces to do or receive work. The actions represented by this spring are always reversible.

The Superficial Observation of Elasticity in Fluids

The elastic or spring like properties of many "heavy bodied" fluids are apparent with manual manipulation. A spectacular material is Silly Putty which is sold as a toy. When rolled into a ball and dropped on the floor it will bounce like a rubber ball. When allowed to rest on a flat surface it will flow slowly across the surface like molasses. Bread dough has the properties of a viscoelastic fluid. When stretched it will recover if released immediately; however, the ability to recover is lost if it is held in the stretched state.

Some industrial printing inks and their ingredients are viscoelastic. The elasticity can best be seen by placing a small portion between the thumb and forefinger (or a spatula and a plate), drawing it into a filament by a rapid separation of the fingers, and then observing the recovery of the filament when the separation of the fingers is decreased slightly (like a rubber band). See Figure 4.

Some materials (such as printing inks) when placed on two rapidly rotating rollers form filaments between the separating surfaces of the rollers. Hull (1951) has reasoned that the elastic properties of the ink causes these filaments to form. His reasoning is that in order for a filament to form it is necessary that there exist an increased resistance to flow in any necked down area; otherwise, the filament would continue to neck down further, for the stresses per unit area are highest at the smallest diameter. Therefore, if a filament is to form, any necking down must cause an increased resistance to flow at the neck. The cause of this increased resistance is an elastic property with which the stress (i.e., the resistance) increases with increasing strain. The increase of stress with increasing strain is characteristic of elastic bodies.†

The behavior of viscoelastic fluids is analogous to the behavior of viscoelastic solids. Materials which have the property of rubberlike elasticity will stretch to a shape which is of uniform thickness, while most other materials neck down relatively rapidly and break (or fracture without necking down). This concept of the relation of elasticity to a materials property of being able to be drawn into a thread has an application in the textile industry. Molten polymers and polymer solutions are extruded into filaments and then drawn down to smaller filaments of uniform diameter.

In the laboratory the elasticity of some fluids can be observed on both the rotating cup, and the cone and plate viscometers. A displacement of the rotating member may show a partial recovery if released immediately. If, instead of releasing, the displacement is held constant for a short time the ability to recover is lost.

The elastic properties of many "heavy" fluids are not sufficient to affect their behavior in flow. For example, corn syrup, honey, and glycerine, follow the laws of Newtonian flow for most practical purposes. However, the elastic properties of many fluids are visible if one looks for them. For those with limited access to fluids which could be viscoelastic, but who still wish to observe viscoelasticity, phlegm is viscoelastic.

FIGURE 4. The elasticity of some viscoelastic fluids can be observed by drawing them into a filament between the thumb and forefinger and then allowing the filament to relax and shorten its length.

Elasticity and the Normal Forces

With further examination of viscoelastic fluids some rather startling effects appear. One of these is the Weissenberg effect; when a rod which extends down through the surface of a viscoelastic fluid is rotated, the fluid will climb the rod to escape from the shear field.

---

†The way in which elasticity aids in the formation of filaments is similar to the way the elasticity of certain bubble films stabilize foams. See for example Rosen (1978, p. 201).

The Weissenberg effect is measurable on a cone and plate viscometer (Walters, 1975). This viscometer consists of a cone with its point resting against a plate. A shear field is formed when the plate is rotated with its center of rotation at the point of the cone. (This design is desirable for the rate of shear is constant for all the material between the cone and the plate.) If manometers are placed along a diameter of the cone, and a viscoelastic fluid is placed between the plate and cone, a rotating of the plate will cause pressures to be generated which will show on the manometers. See Figure 5. The pressure is zero at the edge of the cone and rises to a maximum at the center. A reasonable model for the formation of this pressure is, the shearing of a viscoelastic fluid creates a tension which is parallel to the direction of shear. Since the rate of shear is constant along the radius, the pressure in the manometers P_m is

$$P_m = \int_{r_1}^{R} (P - P_{11}) (1/r) \, dr = (P - P_{11}) \ln (R/r_1) \qquad (1)$$

where R is the outer radius, r_1 is the radius at which the pressure P_m is measured, and $P - P_{11}$ is the tension tangent to circles circumscribed about the point of the cone. This is a somewhat simplified equation, for the tension is assumed to be perfectly parallel to the direction of shear, and the other normal forces are not considered. Equation (1) is in agreement with the experimental results. (Roberts 1954, Katoka 1959)

A more versatile and elaborate form of the cone and plate viscometer is known as the Weissenberg Rheogoinometer. The manometers are replaced by a strain gage which measures the total vertical force against the cone. It has the capability of producing a large range of shear rates, at both constant and rapidly varying shear rates. It is also adaptable for use with molten

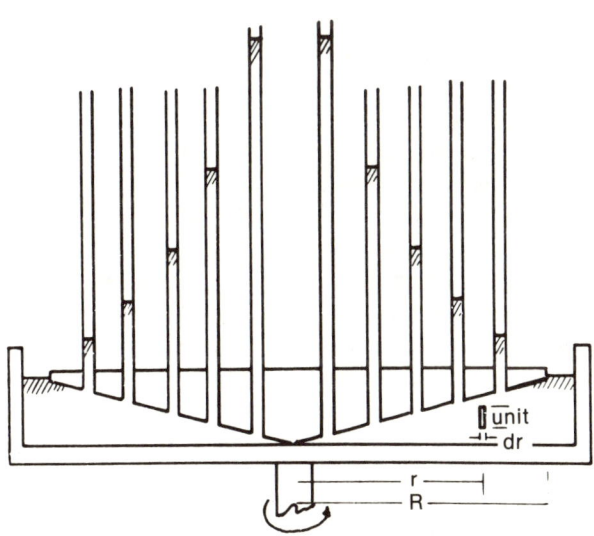

FIGURE 5. The distribution of the pressures in a cone and plate viscometer can be observed by placing manometers along the diameter of the cone.

plastics at high temperatures. However, its use at high shear rates is limited, for viscoelastic materials tend to escape the shear field by flowing to the outer periphery, letting in air. A great deal of work has been done with this instrument on both polymer solutions and molten polymers.

Another example which demonstrates the elastic properties of fluids is the Barus effect or "die swell". Elastic fluids when extruded from a capillary, will swell at the exit. The shearing during the extrusion imposes an elastic deformation on the fluid. The elastic recovery, which occurs after the exit from the end of the capillary, results in the swelling — just as a stretched rubber band swells when allowed to relax. Nielsen (1977) in his chapter "Normal Stresses and Die Swell" gives a list of references on the subject.

Closely allied to the Barus effect is the pressure at the exit of an extrusion nozzle, the relaxation of which presumably produces the Barus effect (as it is equivalent to a relaxation in tension). This pressure is determined by measuring the pressure at various points along the length of the extrusion tube. A plot of the pressures versus their position can be extrapolated to the pressure at the exit. For elastic fluids the extrapolation shows a positive value, while it is close to zero for Newtonian fluids. one of the numerous references on the exit pressure is Funatsu (1970), who also describes the use of photoelastic measurements to confirm the pressure measurements. A commercial instrument is available which measures the exit pressure during extrusion through a slit or a capillary, for the calculation of the elastic properties of molten polymers.

Particles Suspended in Fluids Undergoing Shear

The concepts of stress and stain used in continuum mechanics are helpful in the understanding of the behavior of viscoelastic fluids. However, real fluids are not continuums; they are made up of molecules. One can obtain a gross approximation of the behavior of molecules by considering the action of various types of particles suspended in fluids undergoing shear. Mason and his coworkers have studied these actions in various types of specially constructed apparatus, utilizing motion pictures, still photographs, and visual observation.

Taylor (1934), Rumscheidt and Mason (1961), and Torza, Cox and Mason (1972) have studied the deformation of liquid drops in shear fields. Figure 6 shows tracing of photographs of such drops made at various rates of shear, and at various ratios of viscosities between the suspended and suspending fluids. At zero rate of shear all drops are spherical. At low rates of shear the drops form ellipsoids with their major axes at forty five degrees to the direction of shear. At increasing rates of shear this angle decreases, and in some cases the major axis becomes parallel to the direction of shear. The exact behavior varies with the properties of the two fluids as well as with the rate of shear.

The distortion of the sphere to an ellipsoid indicates that the pressure in the fluid is no longer uniform in all directions. A lower pressure exists in the direction of the major axis of the ellipsoid. The lower pressure is equivalent

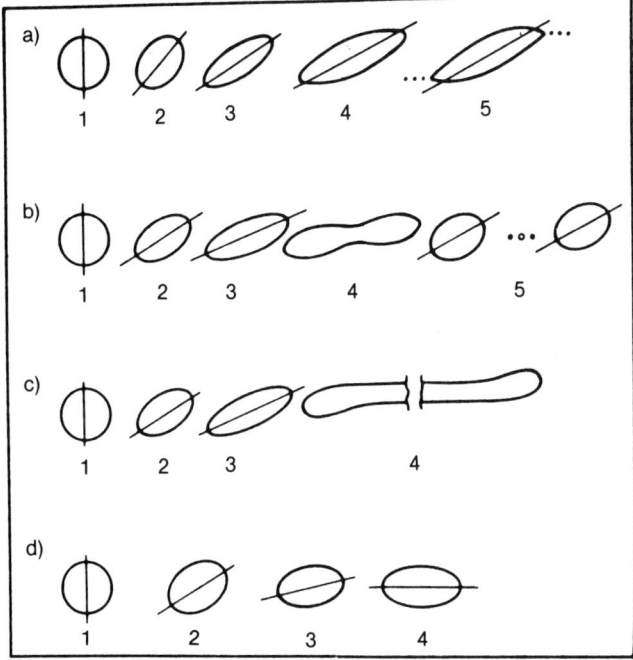

FIGURE 6. Tracings from photographs of drops in shear flow showing the change in drop shape with increasing rates of shear. (a), (b), (c) and (d) have different components in their continuous and/or dispersed phases. From Rumscheid and Mason (1961).

to a tension exerted on the drop in that direction. The forty five degree angle at low rates of shear is in agreement with the tensions and compressions developed by shearing forces described by the methods of continuum mechanics. The development of a tension parallel to the direction of shear is in agreement with the Weissenberg effect shown with the rotating rod and the cone and plate viscometer. The actual details for the deformation of the liquid drops depends on, the properties of the two liquids involved, and on the rate of change of the rate of shear, as well as the rate of shear itself (Torza 1972). Rumscheid and Torza make no mention of the elastic properties of their liquids; however, some of their fluids are silicone oils and Hull (unpublished work) has shown that these oils are not Newtonian at high rates of shear.

The normal forces (P_{11}, P_{22}, P_{33}) should be calculable from the interfacial tension and the shape of the drop, if: the viscosity in the suspended fluid is very low, the inertial effects of the flow around the drop are negligible, and the container walls do not intereferre with the flow (Hull 1969). Consider the cross section of the suspended drop perpendicular to one of its major axes. The surface tension times the length of the periphery at that cross section divided by its area is the pressure exerted by the surface tension on that drop. If the forces caused by the flow of the liquid within the drop are very small (since its viscosity is very low) the pressure within the drop must be equal in all directions. Hence, the differences in pressure exerted along the axes by the surface tension must compensate for the normal forces exerted by the viscoelastic liquid flowing around the drop. The differences in pressure exerted by the surface tension between the three normal directions must then be numerically equal but opposite in sign to the differences between the normal forces. The force exerted by the surface tension is equal to the circumference of the drop times the surface tension. Since the circumference varies with the axes of the ellipsoid the pressure exerted by the surface tension also varies.

Mason's group also studied the behavior of small rods in a shear field (Okagawa 1973, Rumscheidt 1961). The rods rotating about the x_3 axis; however, their rate of angular rotation was not uniform. The rotation was most rapid when the long axis is the rod was perpendicular to the plane of shear and slowest when the rod was in the plane of shear. This caused the suspension to be anisotropic. In a viscoelastic liquid the rods moved to an orientation which gave a minimum resistance to flow; while in a Newtonian liquid at low rates of shear this did not occur.

Notes on Notation

There are many different notations for designating the different normal forces. Numbering the directions of the axes is arbitrary and differs from author to author. Here the axes are numbered as shown in Figure 7. The direction of shearing is (1), the normal to the shearing plates is (2), and the third direction normal to both is (3).

The normal forces are considered to be on an absolute basis so they will be on the same scale as absolute pressure. Hence, placing a body under a moderate tension, will be considered as a lowering of pressure in one direction. The stretching of a rubber band is then a lowering of the pressure on the ends, which allows the atmospheric pressure to squeeze the sides together, elongating the rubber band.

The ellipsoid formed when a small drop of low viscosity fluid is suspended in a shear field is a good model for the visualization of the stress in that shear field. The minimum normal force (or maximum tension) is along the major axis of the ellipsoid. At low rates of shear this is an angle (angle ϕ) to the x_1 axis Figure 8. The major axis of the ellipsoid rotates toward the x_1 direction decreasing the angle ϕ as the rate of shear increases. The P_{11} normal force is parallel to the major axis of the ellipsoid and hence also rotates with the increase in shear rate until it becomes parallel to the shearing faces (the x_1 axis). The normal force P_{22} also rotates with increasing rates of shear, until it becomes perpendicular to the shearing faces. The P_{33} normal force is directed parallel to the x_3 direction. Since this direction is normally open to the atmosphere it is usually equal to the atmospheric pressure. When no information is available as to the angle ϕ it is usually assumed (rather loosely) that P_{11} and P_{22} are completely rotated and that ϕ is equal to zero.

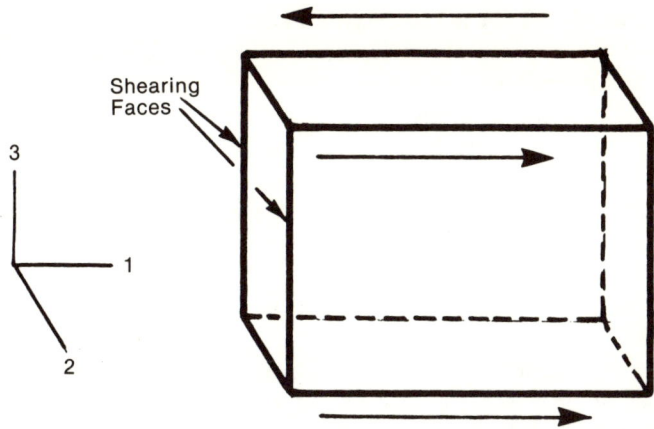

FIGURE 7. The numbering of axes in shear. The direction of shear is (1), normal to the faces is (2), and the direction normal to both is (3). The direction (3) is usually open to the atmosphere and if uniform and there is no flow in that direction is equal to P.

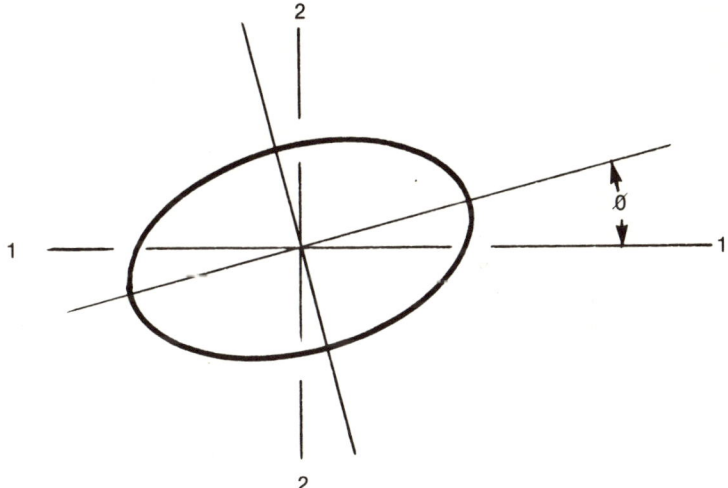

FIGURE 8. A cross section in the (1)-(2) plane of an ellipsoid formed by a drop of liquid suspended in a shear field.

An assumption which is sometimes made in continuum mechanics is that the stress tensor is symetric. This implies that the stress does not produce a rotation. However, all suspended bodies in the shear field of a liquid rotate, and at high rates of shear a turbulence occurs in which vortices are formed. It is more reasonable to consider a model of fluids consisting of finite molecules on which the stress is exerted. Even in a uniform shear field this stress should contain a non-symetrical component. This is more consistent with the rotations and change in the angle ϕ which is observed.

A Molecular Model for a Viscoelastic Fluid

A molecular model for a viscoelastic fluid is a solution or melt which contains an appreciable number of molecules in the form of long chains of atoms. The individual atoms in the chain move about in a near random thermal motion, restricted by possible steric hindrances, by possible limitations on the angles between the atom bonds, and by the atoms being tied together in these chains. These chains assume conformations which are limited random walks, which can in part be described by a probability distribution of the distances between the ends of the chains of atoms. These random walks normally have equal probability distributions in all directions.

When the fluid is undergoing a steady state shear the long chains of atoms assume an average conformation which is much different than that of the "equilibrium" state when the fluid is not undergoing shear. One characteristic of this steady state is that the conformation of the chains varies with direction. When given the opportunity (for example by suddenly stopping the shear) the thermal motion of the chains of atoms will do work to return the conformation of the chains of atoms to their average distribution at the "equilibrium" state. If this results in a bulk flow of the fluid, it is an elastic recovery. If the chains return to their average conformation without the bulk flow but only a rearrangement of the chains, it is an internal relaxation.

The type of chain may vary considerably. It is usually (though not necessarily) a carbon chain. It may vary in length and have various side groups or side chains. Their behavior when undergoing shear would all be modifications of that of the plain chain.

There are other models which can display some of the characteristics of viscoelastic fluids — such as chains of springs or dumbells. The chains of atoms appear to be the most realistic and useful of the models.

A concept of the behavior of chains of atoms in shear fields can be obtained from the work of Mason's group on the behavior of filaments in shear fields (Forgacs 1959, Cox 1971, Okagawa 1975). The behavior of actual chains of atoms are related to the behavior of the filaments they studied. However, *with the chains of atoms the effects of the thermal motion must be much greater than any effects of thermal motions of the filaments* — if the filaments do have any thermal motion. The effects of the thermal motions would be to drive the chains of atoms (or filaments) to more random average positions.

Mason and his students studied the behavior of flexible and non-flexible filaments in shear fields. Figure 9 shows the behavior of filaments of varying flexibility but of the same length. The more flexible filaments conform more to the forces in the shear field. Figure 10 shows the behavior of various length filaments. The short filaments merely bend slightly and rotate. The intermediate length filaments bend sharply back on themselves undergoing 180 degree snake like turns. As the filaments become longer they first form a helix, and then form a complex coil.

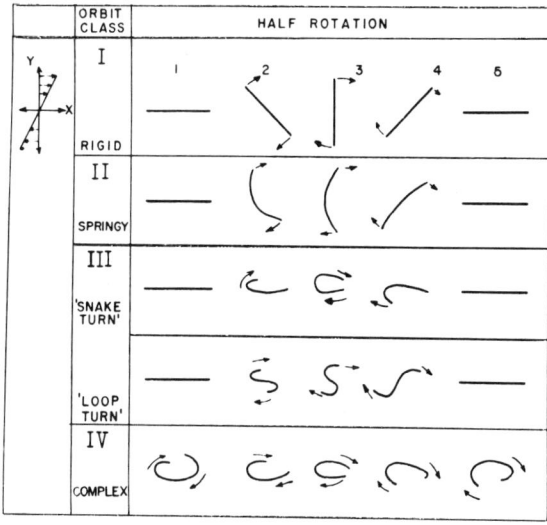

FIGURE 9. The behavior of filaments of the same length but varying flexibility in a shear field. From Forgacs (1958).

Figure 11 (Okagawa 1975) shows tracings made from photographs of a 25 mm. long flexible filament in a silicone oil, at three different rates of shear. Table I shows the average end to end distance of the fiber, in terms of distances parallel to the three axes; h_1 is parallel to the x_1 axis, h_2 to the x_2 axis, and h_3 to the x_3 axis. The axis notation used is that of this book and not that of the original article.

Table I. Measured mean values of end to end distances of filaments suspended in a shear flow. From Okagawa, 1975.

Parameter					
Rate of Shear in sec^{-1}		0.34	0.62	3.12	12.19
h_1	mm.	3.13	3.03	2.93	2.19
h_2	mm.	0.39	0.38	0.25	0.26
h_3	mm.	0.76	1.28	1.09(?)	1.95
V	mm^3	14.84	12.70	9.09	6.96

The following points can be made:

a. A coil is formed which is squashed the most in the x_2 direction, and h_2 appears to decrease slightly with increased rates of shear.

b. The coil is spread out the most in the x_1 direction, though this spread appears to decrease slightly with increased rates of shear.

c. The volume occupied by the coil decreases with increased rates of shear. This apparently reflects the formation of a tighter coil at increased rates of shear.

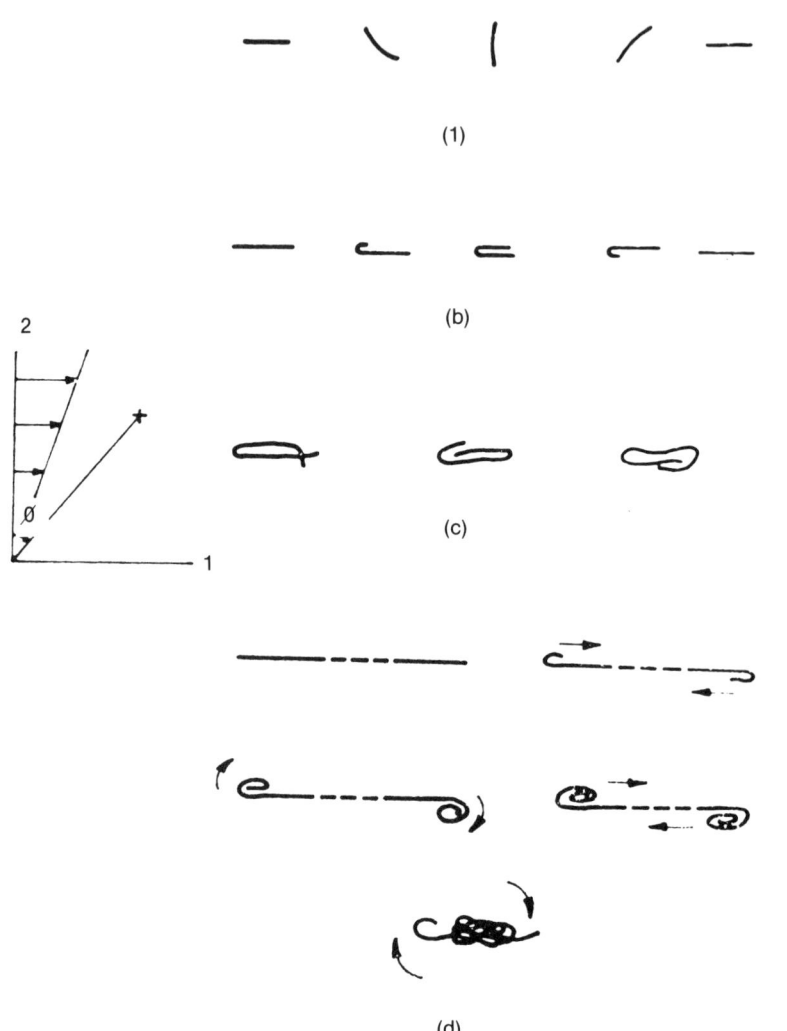

FIGURE 10. The behavior of filaments of varying length in a shear field. From Cox (1971).

One would expect real differences between the conformations of filaments and the conformations of molecules made of carbon chains in shear fields. However, the pictures of the filaments and the rods provide a basis from which a model for the behavior of the carbon chains can be constructed. The prime differences between the particles photographed are expected to be: the carbon chains would have much more thermal motion, and the carbon chains are probably much more flexible. These properties can be imposed on the behavior of the filaments shown in Figure 11.

The effect of a shear field on carbon chains can be described in terms of: their average end to end distance, the orientation of that end to end

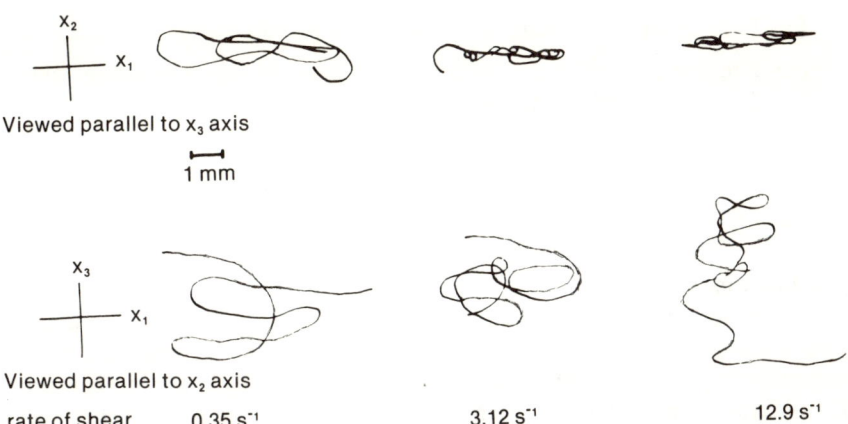

FIGURE 11. Tracings made from photographs of a 25 mm. filament suspended in a shear field at three different rates of shear. The photographs are from Okagawa (1975) and the tracings were made by Hull. See Figure 7 for axes designation. "Viewed parallel to the X_3 axis" means the line of sight is parallel to the X_3 axis.

distance with respect to the axes, and a coiling of the molecules. In a field in which the rate of shear is zero their average end to end distance would be the same in all directions; the average orientation would be the same in all directions; and their would be no coiling of the chains. In other words their distribution would be the same in all directions; and their would be no coiling of the chains. Their distribution would be symmetrical.

In a shear field the distribution of the carbon chains would no longer be symmetrical. A reasonable model would be: (a) their average end to end distance is greater than the average in the x_1 direction and less than the average in the x_2 direction, (b) there would be on the average more orientation in the x_1 direction than in the x_2 direction, and (c) there would be a coiling of the molecules which would be especially evident at the higher rates of shear. The thermal motion of the molecules would generate internal forces which tend to drive the molecules back to their (average) symmetrical positions. These internal forces are evident in the tension generated parallel to the x_1 axis and in the compressional forces exerted against the shearing faces.

When a viscoelastic fluid is allowed to relax, the molecules resume their symmetrical conformation. This may take place by either a bulk movement of the fluid (as in the Barus effect or the recovery of a cone and plate viscometer) or by a movement of the molecules themselves — an internal relaxation.

Justification of Thermodynamic Methods for the Steady State in Shear

The proof that theromdynamics (similar to those of "equilibrium" thermodynamics) can be applied to viscoelastic fluids in the steady state is based on the following three points:

(a) The equations of state are valid in the steady state.

(b) The reversible and irreversible processes are separable over definable steady state paths.

(c) The necessary theromdynamic functions are definable for the steady state.

Each of these points will be discussed in turn.

The Equations of State

Consider a viscoelastic fluid held between two faces which exert a constant stress X in shear on the fluid. A surface of the fluid is exposed to the surrounding pressure P. The shearing faces are held at constant temperature T, and these faces have a high thermal conductivity so that the heat generated by the viscous component of the flow is conducted away rapidly. But, no matter how efficiently the heat is conducted away the temperature across the fluid (from shearing face to shearing face) will not be uniform. However, the smaller the distance between the faces the smaller the temperature differences and as this distance approaches zero the temperature differences approach zero. The equation of state,†

$$B = f(P, T, X) \qquad (2)$$

applies to the limiting condition reached as the distance between the faces approaches zero.

This limiting condition is within the tradition of classical thermodynamics. For example, the Carnot cycle is a cycle for which two quantities must approach zero: (a) friction must approach zero, (b) the thermal resistance for the flow of heat between the cycling body and the heat sink and between the heat source must approach zero. These two assumptions for the Carnot cycle are limits just as those stated above for the viscoelastic fluid in the steady state.

If the equation of state is valid for the state, (P_1, T_1, X_1) the properties of the thermodynamic body (the "B"s in equation (2)) must be the same no matter what path is taken to that state. If this were not true in an experimental situation one would look for a reason, such as, either the correct steady state had not been reached or an unexpected chemical reaction has occurred.

Hysteresis occurs in solids and often interfers with the attainment of an equilibrium state. Hysteresis is much less likely to occur in fluids, for the molecules are much freer to move and conform to the forces involved. Problems in the attainment of the duplication in the measurement of properties in a supposed steady state are likely to be caused by polymer stress reactions (Casale, 1979) or poor temperature control.

---

†Since the molecular weight may be unknown B represents a dependent quantity in terms of per unit weight.

The Separability of the Reversible and Irreversible Processes

The separability of the reversible and irreversible processes is implied in the modified Maxwell model with a kinetic energy spring, a molecular force spring, and a dashpot as shown in Figure 3. This implied separability is correct for real viscoelatic fluids. The separability of the two processes can be demonstrated in a way similar to that shown to be valid for solids.

When a viscoelastic fluid is sheared in the steady state (at constant P, T, X) and all its properties remain constant though heat continues to be generated in that body and flows out. Since this a a visco*elastic* fluid it is implied that some ability to do work is stored in the fluid. This "elastic energy" is constant in the steady state, though the amount of irreversible heat generated is not. It is necessary to show that the reversible and irreversible processes are completely separable in this steady state.

Consider now that the shearing force is slowly increased from X_1 to X_2, keeping the temperature and pressure constant. Each point on that path is a steady state position. Although irreversible work continues to be done and its heat equivalent removed, there will be energy changes within the viscoelastic body which are completely separate. Between each adjacent point on the steady state path there will be an amount of work done on the body (above the irreversible work) which is equal to an incremental change in internal energy dE plus an incremental change in the entropy with an energy equivalent of $-T\, dS$, plus any incremental work done on the atmosphere $P\, dV$. The sum of all these incremental changes will be,

$$\Delta E^{PT} - T\, \Delta S^{PT} + T\, \Delta V^{PT} = \Delta G^{PT} \tag{3}$$

which is the change in elastic (i.e., reversible) energy within the body in going from state (1) to state (2). However, no matter what path is taken from state (1) to state (2) the difference in the stored elastic energy is the same, for, it is a B in the equation of state (2).

If the path is reversed going from state (2) to state (1) the body must release an amount of energy equal to that previously absorbed over that route and equal to that of equation (3). However, that released energy would not be completely available as external work as it would be for elastic solids. A portion, if not all the elastic energy released, would be used to temporarily maintain the irreversible processes.

It is possible to construct pseudo-adiabatic paths for viscoelastic fluids analogous to those described for rubberlike solids. Consider that the force X_1 is changed to X_2 very rapidly and that the irreversible heat generated during the resultant increase in deformation is either negligible or removed by some intelligent machine. The temperature will rise from T_1 to T_2 just as it does when rubber is stretched adiabatically. The viscoelastic fluid is then held at the steady state P, T_2, X_2. The entropy of the viscoelastic fluid is constant, since only the heat generated by the irreversible processes has been removed.

It is now possible to combine the isothermal and the pseudo-adiabatic paths to create a pseudo-Carnot cycle. There would of course be little or no net work available from such a pseudo-Carnot cycle for the losses from the irreversible processes which take place at the same time would be high. This cycle has no practical value other than to emphasize the correspondence between steady state thermodynamics and "equilibrium" thermodynamics.

In equilibrium thermodynamics it is possible for a body to go from any equilibrium state to any other equilibrium state over equilibrium paths. With a gas not more than two paths are needed — one adiabatic and one isothermal. These equilibrium paths are the basis for the calculation of the changes in entropy and other thermodynamic properties between these states.

Steady state paths can also be described which are equivalent to the isothermal and adiabatic paths of equilibrium thermodynamics. It has just been shown that the reversible component of the processes on these paths is completely separable from the irreversible components. Thermodynamic equations can consequently be written for these reversible processes which are analogous to those of "equilibrium" thermodynamics over the reversible paths.

The Definability of the Theromdynamic Functions

The third requirement is that the thermodynamic functions (such as A, H, G, and GG) be valid for viscoelastic fluids in the steady state. For this to be true it is only necessary that their components as included in their definitions be real and valid. These components are, E, P, V, T, S, X, and Y. The validity of P, V, T, and X for a body undergoing steady state shear is self evident. E is the sum of the kinetic and intermolecular potential energies (plus the energy within the atoms). As such it has the same validity as in equilibrium thermodynamics. Y represents the recoverable deformation in shear — if there were no internal relaxation during that recovery. This should be obtainable as an extrapolation from measurements of actual rates of actual recoveries and actual rates of force decay, after shearing has stopped. This may be difficult to measure though it is a valid quantity.

This leaves entropy S, which has two different definitions. The *statistical* definitions of entropy differences is $\Delta S = k \ln W$. For a viscoelastic fluid in the steady state the probability W, that the molecules would (when at zero rate of shear) attain the same conformation, by chance and chance alone, that they have when undergoing shear. This is a reasonable definition for W, and this in turn validates the associated concept of entropy.

The justification of the *thermal* concept of entropy depends on the separability of the reversible and irreversible processes. Consider a viscoelastic fluid under going shear in the steady state at P_1, T_1, X_1. The force in shear is then increased so that the steady state is P_1, T_1, X_2. The heat transferred from the fluid in going from X_1 to X_2 is derived from two separable processes — the irreversible and the reversible. The change in the entropy is equivalent to that in conventional thermodynamics — the reversible heat absorbed divided by the temperature. A more detailed description and comparison of the processes involved follows.

Entropy Changes in a Viscoelastic Fluid Between Steady States

Consider the compression of a real gas at constant temperature. The work of an incremental amount of compression is $P\,dV$, which is converted to thermal energy in the amount of δQ_1 which is absorbed by a heat sink. In addition to the external work the attraction of the molecules to each other aids in the compression and does work in the amount of,

$$dE = Pi\,dV = \delta Q_2 \qquad (4)$$

which is converted to heat in the amount of δQ_2 which is also absorbed by the heat sink. Pi is the force of attraction of the molecules to each other expressed in terms of pressure. See equations VII-13,14. The sum of both incremental heats is absorbed by the heat sink giving a change in entropy in the gas of,

$$(\delta Q_1 + \delta Q_2)/T = dS_{re} \qquad (5)$$

This change in entropy of the gas is completely reversible.

Now consider the analogous process by which a change in the state of a viscoelastic fluid is generated by a force exerted in shear X. During the time the force X is being established both irreversible work and reversible work is being done on the body. Once the steady state has been reached all the work done on the body is irreversible work which is converted to heat and absorbed by the heat sink. In reaching the steady state three separate molecular processes are involved. As incremental amounts there are:

(a) The work done by X against the thermal motion of the molecules changing their position to one which is less probable on the average. This work is equal to,

$$X\,dY = \delta Q_1 \qquad (6)$$

since all this work is converted to heat δQ_1 which is absorbed by the heat sink. dY is the incremental elastic displacement (i.e., that which would be recoverable if there were no internal molecular relaxation).

(b) A displacement caused by X is aided (or opposed) by the attraction of the molecules to each other. This internal attraction does work against the thermal motion of the molecules and is hence converted to heat in the amount of δQ_2 which is equal to,

$$dE = Xi\,dY = \delta Q_2 \qquad (7)$$

where the attraction of the molecules to each other acts as an internal force in shear. This is analogous to the internal tension of equation VII-18.

(c) If there is a change in volume of the viscoelastic fluid reversible work will be done by or against the thermal motion of the molecules by the pressure P. This utilizes heat (or conversely generates heat) and is equal to,

$$P\,dV = \delta Q_3 \qquad (8)$$

All the heats generated by the reversible processes above are absorbed by the heat sink, giving an entropy change of the viscoelastic fluid of

$$(\delta Q_1 + \delta Q_2 + \delta Q_3)/T = dS_{re} \tag{9}$$

This is analogous to the molecular processes associated with the change in entropy of a gas, but even more closely associated to those which occur with the stretching of a rubber body at constant temperature and pressure. This change in entropy is also reversible.

The Significance and Utility of the Thermodynamic Functions

Once the validity of the thermodynamic functions has been established, their significance and utility can be investigated. As might be expected their uses are similar to their uses with viscoelastic solids, except that they apply to balance in the steady state rather than to equilibrium. The following are examples.

The criterion for the equilibrium of a solid body depends on which model is selected as valid for the system. For a solid body at constant P, T, and X the criterion of equilibrium is a minimum GG. See VII-23. The same criterion can be applied to balance in the steady state; however, the criterion is for the limiting condition of an infinitesimally thin film in the steady state. The same proof applies to a body in the steady state as it does to an elastic solid. The ability of a body to do work in that particular system (i.e., it's free energy) must be at a minimum under the restraints imposed.

The other criterion for equilibrium are applicable as tests for balance in the steady state in the same way. For example, from statement VII-29, for any given E, V, and Y the criterion for balance in the steady state is that the entropy S be a maximum. In "equilibrium" thermodynamics these conditions form what is referred to as an isolated system. In the steady state this system could not be isolated, as energy would be constantly flowing through it. However, the criterion is still valid, even though such a system would be difficult to construct.

With a solid elastic body it has been shown that the two variables P and T can be selected as the only independent variables of state; however, then the requirement for equilibrium of a minimum G implies that no elastic deformation be present. The same option of two independent variables of state for a viscoelastic fluid in the steady state may be assumed; but just as in the viscoelastic solid, a minimum G or no elastic deformation is the criterion of balance. With two independent variables of state, balance in the steady state and equilibrium merge to the same state.

The ΔG^{PT} of a viscoelastic fluid in the steady state is a Gibbs free energy, and this has all the implications of the classical concepts of free energy. At the same time, though it is the maximum available work (at constant P and T) a mechanism must be present by which that work can be done. With free energy available from chemical reactions the device for converting that free energy into electrical work is a battery. One can conceive of a galvanic cell made with two electrodes of the same metal one with strain and one strain free. However, cells of this type are difficult to demonstrate for voltages depend on the actual reactions which occur at the electrodes. For example such an electrode can usually also act as a hydrogen electrode. Strained polymers cannot be used to make galvanic cells for battery-like reactions are not known at present.

With elastomeric solids the path by which the elastic energy (the free energy ΔG^{PT}) can be recovered as to allow the solid body to do work against an outside force as it returns to its original shape. Viscoelastic fluids have a unique way of converting elastic energy (ΔG^{PT}) to work which is not available to elastic solids. Since the molecules in the viscoelastic fluid are not held together into a network by cross-links *they can flow to other areas which are not undergoing shear* and in so doing, do work. The tendency of viscoelastic fluids to escape to areas which are not undergoing shear is evident in several phenomena. However, it is best measured by means of the pressures developed in ports to areas undergoing shear.

Consider a viscoelastic body undergoing a uniform steady state shear between two parallel plates as in Figure 12. One of these plates is in a fixed position and there is a small opening in that face designated as (E) in the figure. Within the volume (ABCD) material is being added and lost at the same rate so that the material within the volume ABCD is constant and can be considered a thermodynamic body. Equations for the energy balance of that body can be written.

First consider that P, T, and X are constant and that only the reversible actions on the body ABCD be considered. The addition and removal of material to this body sum to zero, hence they have no effect. The conditions for balance in the steady state are equivalent to those of the solid body with the same restrictions. From equation VII-18 this is a minimum GG^{PTX}.

Second, consider the relationship between the fluid in the hole E and the material being sheared. There is a difference in the Gibbs free energy ΔG^{PT} equal to the elastic energy of deformation. This difference in ΔG^{PT} will cause the fluid to attempt to flow to the area of lower free energy, creating a pressure in the hole E.

In terms of molecular models the pressure is the result of the three terms in the Gibbs free energy. These are,

(a) The entropy effect from the term $-T \Delta S^{PT}$ which is the result of the thermal motion of the molecules attempting to drive them to a position where their conformation is equal and symetrical in all directions and equal to the conformation attained in a field not subjected to shear.

(b) The intermolecular force effect from the term ΔE^{PT}. In rubberlike elasticity this effect usually opposes the effect in (a) above.

(c) The effect of any change in volume $P \Delta V^{PT}$. The work associated with any change in volume is of course done by or against the pressure caused by the thermal motion of the molecules.

The actual pressure developed in the hole E depends on three quantities:

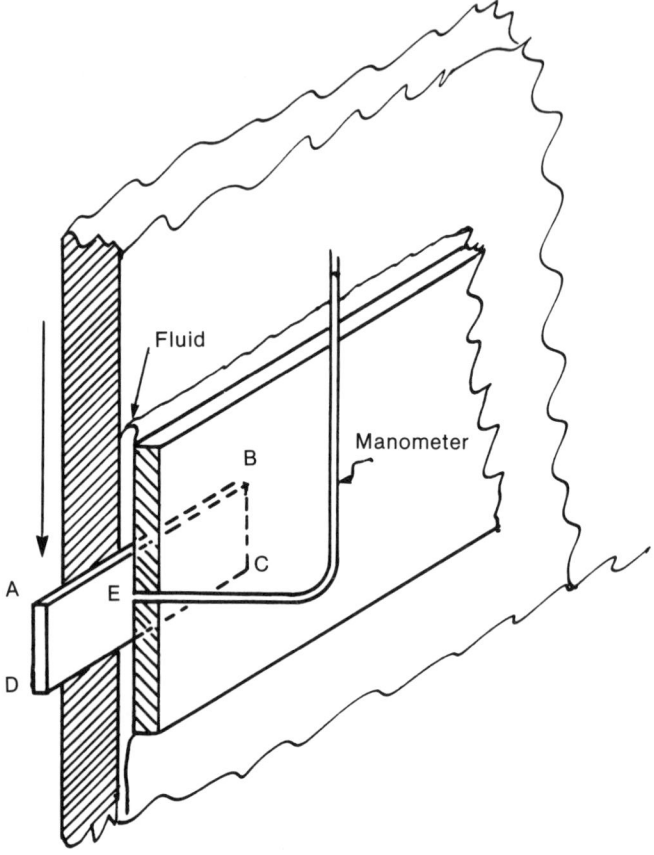

FIGURE 12. Diagram showing a fixed volume within a sheared fluid. The shearing surfaces are shown in cross section. A manometer is placed in the fixed plate. For a similar example see Figure 14.

(a) The magnitude of the Gibbs free energy ΔG^{PT} which is the maximum reversible work attainable.

(b) The mechanics of the process of molecular relaxation which may limit the pressure exerted to less than the possible maximum and which must account for the change in pressure with the direction.

(c) A component of the pressure associated with the viscous component of the flow which will be designated P_{ir}.

Since $P_{re}V$ is the maximum reversible work obtainable from the passage of the viscoelastic fluid through hole E, we can write,

$$(P_{re}) V \leq \Delta G^{PT} = \int_0^Y X\, dy \qquad (10)$$

where P_{re} is the reversible component of the pressure developed in the hole

E, and *dy* is the increment of the reversible elastic displacement in shear.

The total pressure at the exit is the sum of the reversible and irreversible pressures or

$$P = P_{ir} + P_{re} \qquad (11)$$

where the pressure P in this equation refers to the incremental pressure above atmospheric.

In practice P_{ir} can be obtained by measuring the total pressure against the wall at various positions along an extrusion tube. P_{ir} becomes zero at the exit while P_{re} does not. Hence, P_{re} can be obtained by extrapolation of the total pressure to the exit.

The thermodynamic partial differential equations developed in Chapter VII for solids apply to viscoelastic fluids in the steady state as well as to equilibrium or static balance. This applies to the Maxwell relations as well as those which contain the thermodynamic functions. They are of course limited to conditions such that the heat generated does not produce an appreciable temperature gradient.

There is also an internal shearing force just as there is an internal pressure in gases and an internal tension when stretching rubber. See equations VII-13, 18 and VIII-7.

$$Xi = (\partial E/\partial Y)_{TV} = T(\partial X/\partial T)_{VY} - X = T(\partial S/\partial Y)_{TV} - X \qquad (12)$$

The Xi is an internal shearing force which for most rubberlike materials aids the external stress in shear in changing the conformation of the molecules from completely random to some ordered or partially ordered state. The thermal motion of the molecules drives the molecules to more random positions, while the intermolecular forces drive the molecules to positions in which the intermolecular potential energy is smaller.

The result of these three actions is to force the molecules to a conformation which gives a minimum free energy (ΔGG^{PTX} for the conditions of constant pressure, temperature, and stress).

A Thermal Measurement on a Viscoelastic Fluid

Temperature measurements can be useful in the interpretation of the behavior of viscoelastic fluids. For example, temperature measurements were made in an experiment similar to those conducted by Good (1971). See Chapter VII. Instead of measuring the temperature of the adhesive as adhesive tape was removed, the temperature of a fluid was measured as it was temporarily subjected to a high rate of shear. A small amount of polymerized linseed oil (known in the trade as varnish number eight) was placed on a glass plate. A thermocouple made of very small wires (0.003 inches in diameter) was placed in the oil, and a thin flexible ribbon of cellulose acetate placed on top. The thermocouple was connected to an oscilloscope so that rapid changes in temperature could be recorded. Figure 13 shows an example of the temperatures recorded, when the tape was rapidly stripped from the glass (Hull 1967). The interpretation is the same as

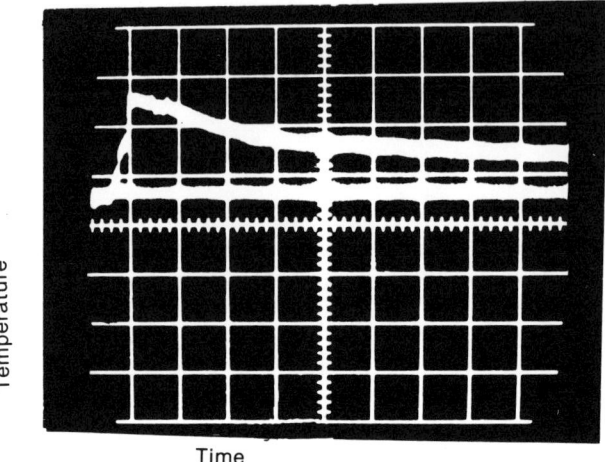

FIGURE 13. The temperature changes which occur in a polymerized oil as it is sheared very rapidly and then allowed to relax. The temperature rise is about 8 °F and the time interval between lines on chart is 0.005 sec. Analogous to Figure 6 Chapter VII.

that given for the temperature changes in the adhesive of the tape. See Figure 8 in Chapter VII.

The very rapid rise is caused by the sum of the external work and the internal work being converted to heat. The external work includes both the reversible and irreversible work. With the rupture of the oil film the stresses in the film relax rapidly. The first fall in temperature is cause by the internal work done by the thermal motions of the molecules against the intermolecular forces. This is the work done against the internal stress of equation (12). This internal work is equal to the energy equivalent of this first temperature drop. Where the temperature change is relatively slow the changes are considered to be due to conduction and radiation. The residual temperature rise (before the effects of radiation and conduction) is a measure of the total external work done during the stripping of the tape from the glass. The reversible component of the external work has become irreversible by the relaxation processes. In this experiment the external work happens to be approximately equal to the internal work.

This experiment is primarily a confirmation of Good's experiment, but it does confirm that transient temperature measurements can be useful in the study of the rheology of viscoelastic fluids. These are similar to those which occur in the deformation of rubber. With instruments which are more sensitive to temperature and possibly smaller thermocouple wires (to decrease the temperature response time of the thermocouple) this type of measurement would provide useful thermodynamic information on the behavior of viscoelastic fluids.

The Two Errors of Garner Nissan and Wood

Garner, Nissan, and Wood were pioneers in the study of viscoelastic fluids. They were aware of, the Weissenberg effect, the Barus effect, the high pressure drop at the entrance to a capillary, and other effects associated with the flow of viscoelastic fluids. They correctly attributed these effects to the elastic properties of these fluids. In their earlier papers they stated that the elastic energy of these fluids was a Gibbs (or Helmholtz) free energy (Garner 1946, 1947, 1949, 1950); however, they made an error in the thermodynamic interpretation of their data and compounded their error by retracting their original statement — that the elastic energy was a Gibbs free energy. An outline of their experiments and their errors is informative.

The constructed a rotating bob type device and measured a normal pressure generated in the device. The details of their device and their method of taking measurements will not be described here. Their normal pressure was measured at two rotational speeds and at three temperatures. Their data is shown in Table IX-II. The normal pressure was measured in terms of centimeters of head and equated to the Gibbs free energy — expressed in the same units. The change in internal energy associated with the shearing action was calculated from the two equations,

$$(\partial \Delta G/\partial T)_p = - \Delta S \tag{13}$$

$$\Delta G = \Delta E - T \Delta S + P \Delta V \tag{14}$$

From these equations they calculated very high values of ΔE (from 27 to 148 cm. of head) which were obviously wrong. Their retraction was based on these calculations.

Table IX-II. The measurements of pressure head at two speeds of rotation and three temperatures. From Garner (1950)

rpm	Temperature	ΔG as cm of head (corrected)		
		8.0°C	20.0°C	30.0°C
1000		6.1	4.9	2.8
1200		6.4	5.6	3.2

The error in their calculations is an omission of the condition if the rate of shear being constant for each of the three measurements at different temperatures. In other words they used two-variable thermodynamics when they should have used three-variable thermodynamics. The correct partial derivative of ΔG^{PT} with respect to T should be taken with a notation that the rate of shear ($\dot{x}$) is constant (Hull 1961, 1969).

$$(\partial \Delta G^{PT}/\partial T)_{P\dot{x}} = (\partial \Delta E^{PT}/\partial T)_{P\dot{x}} - T(\partial \Delta S^{PT}/\partial T)_{P\dot{x}} - \Delta S^{PT} \\ + P(\partial \Delta V^{PT}/\partial T)_{P\dot{x}} \tag{15}$$

With two-variable thermodynamics the first, second and fourth terms sum to zero giving equation (13). This is not in equation (15).

When ΔG^{PT} is plotted against temperature it reaches a value of zero just above 30 °C. Then by an iterative process the values of $(\partial \Delta E/\partial T)_{P\dot{x}}$ can be calculated. (The volume effects are considered negligible.) The values obtained are given in Table IX-III. From this table we can say that the internal energy component is approximately two percent of the free energy. Hence the major component of the free energy for his material is the $T \Delta S^{PT}$ term. Their material shows elastic behavior close to that of an ideal rubber, and hence is quite different than the materials used to obtain the data in Figures VII-8 and IX-13.

Table IX-III. The calculated values from equation (15)

$(\partial \Delta E/\partial T)_{P\dot{x}}$ (in cm. of head)

rpm	Temperature	8.0 °C	20.0 °C	30 °C
1000		−0.011	−0.011	−0.016
1200		−0.013	−0.001	−0.019

An emphasis has previously been made of the importance of noting which conditions are held constant in the differential equations. The above errors of Garner, Nissan, and Wood would probably have been avoided if this had been done in the above example.

Comments on Some Viscometric Flows

Flow through a capillary has been long studied. It is an ideal means for the determination of the viscosity of Newtonian fluids at low rates of shear. There are complications in the flow of non-Newtonian fluids through capillaries which can be both informative, and at the same time a hindrance in the measurement of the properties of such fluids.

According to the theory of capillary flow the stress in shear in a capillary varies from zero at the center to a maximum at the wall. At a distance r from the center of the capillary the shear stress is $\Delta P \pi r / 2\ell$, where ΔP is the pressure drop along a uniform section of the capillary and ℓ is the length of the uniform capillary over which that pressure drop occurs. Both the elastic energy (ΔG^{PT}) and the normal pressure perpendicular to the wall are a function of shear stress, so these should also vary from zero at the center of the capillary to a maximum at the wall. Such conditions should create an unstable flow, and there is evidence that this unstable flow does develop.

Bagley (1956) and Tordella (1957) studied the flow of molten polyethylene by suspending colored particles in the molten polyethylene, and observing their behavior at the entrance to a capillary. At low rates of flow the particles flowed in a funnel like formation toward the entrance. Portions stayed in corners (away from the entrance) and did not flow. When the rate of flow through the capillary was increased there came a point at which an unstable condition developed. The material in the funnel like pattern broke toward the corners, forcing the material in the corners into the capillary. This stopped and then repeated, oscillating back and forth. The oscillations became more violent as the rate of flow increased. At the same time that the oscillations started at the entrance, the extrudate lost its smoothness, and at the higher rates of shear the extrudate became very irregular in shape and finally broke into pieces as it left the capillary. This latter phenomenon is called melt fracture. These phenomena can be explained in terms of the Gibbs free energy. Viscoelastic material when undergoing shear trends to flow to areas in which its free energy is lower. Hence, any pattern of flow is unstable if viscoelastic material undergoing relatively high shear rates borders other material in which the shear rate is much lower. The uneven shapes of the extrudate are caused by the residual stresses generated during the oscillations at the entrance to the capillary. Similar experiments have been reported by Goto (1978) and many others.

A related unstability is present at the higher rates of shear in both the cone and plate viscometer (Han, 1974) and the rotating cup viscometer. The flow is stable in these instruments at the lower rates of shear, but the rate of shear is limited. There comes a point at which the viscoelastic fluid will not stay in the shear field. The fluid flows to the open edges where it can escape the shear field. Air replaces the fluid causing unstable and invalid readings. At low rates of shear the fluids apparently exist in the metastable state, similar to the states of supersaturation, supercooling and superheating of fluids.

Another demonstration of the flow of viscoelastic fluids to areas of lower shear rate has been reported by Tanner (1970). Fluids are allowed to flow in inclined open channels. With Newtonian fluids the surface of the fluids in the channels is flat. With viscoelastic fluids the surface of the fluids is raised in the center. The rise is caused by the diversion of the fluid from the edges where the shear rate is higher to the center where there is a negligible shear rate. Thus this phenomena is evident at even very low rates of shear.

When a viscoelastic fluid contains two components such as polymer (component 2) dissolved in a solvent (component 1) the polymer could impart the elastic properties to the solution. Any free energy imparted by a shear field would then be primarily contained in the molecules of the polymer. When there is a gradient in the rate of shear of such a viscoelastic solution there will also be a gradient in the free energy but the free energy will be primarily in the polymer component of the solution. This is expected to cause the polymer to migrate from the areas of high shear rate, creating a concentration gradient. A discussion of the thermodynamics of gradients is in the next chapter.

The cone and plate viscometer and the rotating cup viscometer are designed to subject their test samples to uniform rates of shear — thus eliminating shear gradients. However, their maximum rates of shear are limited. The band viscometer is also designed to subject a sample to uniform rates of shear. Much higher shear rates can be attained (Hull, 1952,

Wachholtz, 1940, 1941). See Figure 14. A plastic tape is placed between two flat blocks shimmed apart to give approximately 0.002 inches clearance on both sides of the tape. A small amount of the fluid to be tested is placed in a shallow well at the top of the block. The tape is pulled between the blocks by means of weights. The velocity of the tape is determined by measuring the time for the tape to travel a fixed distance (after a constant speed is reached). This design permits viscous fluids to be measured at high rates of shear. Temperature control is good as the fluid film is very thin and the temperature of the blocks is controlled.

Hull (1969) has investigated the pressure developed at the face of the blocks as the band is pulled through. The presence of this pressure is indicated by a flow of fluid into the space between the blocks at the side of the band and then back up into the wall as the tape is pulled through. It was shown that a high pressure existed at a hole in the face of the block; however, the magnitude of the pressure was beyond the range of the manometer used. It is presumed that the high pressure is a result of both viscous flow and the normal pressures resulting from the free energy of the viscoelastic fluid. The band viscometer is useful for the study of the flow of viscous fluids at high rates of shear. The main problems with the band viscometer are: pressure at the sides of the tape cause fluid to circulate to the side of the tape then back up to the top, and it is difficult to get a band of the required uniformity in thickness.

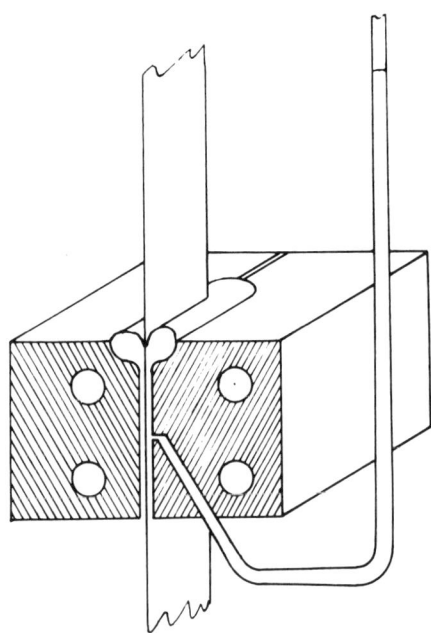

FIGURE 14. Diagram of the cross section of a band viscometer with a hole in one face to measure pressure. The pressure was too large to measure with the manometer shown. From Hull (1952).

The rod viscometer has a geometry which does away with the effects of the flow at the edges of the band with the band viscometer. It consists of a block with a round hole and a rod which fits within the hole leaving about 0.002 inches clearance. The fluid to be tested is placed in a small depression at the entrance to the hole in the block. Weights are used to push the rod through the hole, carrying the fluid with it. The velocity of the rod is calculated from the time it takes the rod to fall a given distance (after it has reached a constant velocity). The weights may be either placed on the top of the rod or attached to the bottom of the rod. When the weights are on the top of the rod, the rod is kept centered and vertical by the large forces generated between the rod and the faces of the hole. The rod viscometer is a simple device which is used for the routine control of the apparent viscoeity of viscous materials such as printing inks. (Hull 1963, 1971, 1966, Tollenaar 1955)

No device used for the study of the flow of fluids is without faults. The capillary, the band, and the rod viscometers all have entrance and exit effects. The entrance and exit effects can be eliminated by the use of rotating type devices such as the rotating cup or the cone and plate viscometers; however, these are so constructed that they have an edge effect where the shear rate drops to zero. One effect of this is that the fluid flows from areas of high shear rate to areas of lower shear rate. This creates a circulation which is probably minor at low shear rates but which draws in air at the higher shear rates. This limits the range over which tests can be made. The curvature of the shear field characteristic of these devices can be used to aid in the measurement of the normal forces.

With rotating cup type viscometers the lower stress per unit area is next to the wall of the cup. When polymer solutions are used this difference in stress will tend to drive the polymer component of the solution toward the walls of the cup (away from the bob). On the other hand the shearing action in the *curved field* will generate a tension in the polymer component of the solution which should tend to drive the polymer component away from the wall of the cup. Any actual effect will depend on which of the actions predominates.

The annular-flow-type instrument is relatively new, and some interesting measurements have been obtained from it. Figure 15 shows its design. An extrusion tube contains a solid rod. Pressure measuring devices are opposite each other on the outer wall and the inner rod. When material is extruded through the annulus the pressure difference between the outside wall and the inside wall (i.e., on the rod) is measured. Okubo and Hori (1980) report the pressure is greater on the outside wall than on the inside wall. This indicates the material between the walls is attempting to expand in the x_3 direction (see Figure 7 for the meaning of the directions); that is, it acts as if it were under compression. This is the opposite effect to that obtained in the cone and plate viscometer where the pressure increases toward the center of the curved shear field indicating in this case a tension in

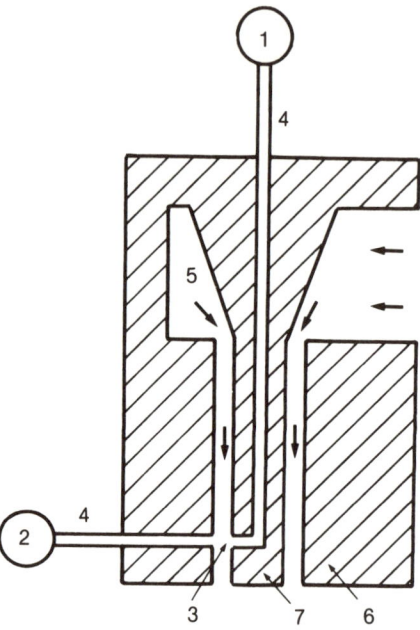

FIGURE 15. Schematic Drawing of an annular-flow instrument (after Okubo and Hori, 1980). The arrows show the direction of flow of the extruded material; 1, pressure transducer for inner pressure; 2, pressure transducer for outer pressure; 3, pressure holes; 4, pipes leading from pressure holes to transducer; 5, reservoir; 6, outer cylinder; 7, inner cylinder.

the x_1 direction. There is relatively little data available from annular-flow-type instruments so there is no assurance that the above relationship holds for other materials or other conditions.

Chapter X

Thermodynamics of the Steady State in the Presence of Gradients

The General Problem of Thermodynamics and Gradients

There has been a great deal written about thermodynamics in the presence of gradients. It is generally called the thermodynamics of irreversible processes. The name correctly implies the emphasis of the usual approach — the irreversible processes. Here the emphasis is different — the *properties* of the bodies in which irreversible processes are occurring, and on the *reversible actions* on these bodies which result from the irreversible processes. The book with the closest viewpoint to that in this book is Denbigh *Thermodynamics of the Steady State* (1951), though, the viewpoint taken here is somewhat different. Denbigh is also used as a primary reference for the material in this chapter.

The three points used for the development and justification of the thermodynamics of viscoelastic fluids in the steady state can also be used here. These points are,

(a) The equations of state are valid for the steady state in the presence of gradients.

(b) The reversible and irreversible processes are separable over definable steady state paths.

(c) The necessary thermodynamic functions are definable for the steady state in the presence of gradients. Each of these points will be discussed in turn.

The Equations of State

The equation of state takes the form of equation II-12

$$B = (P, T, X_3 \ldots X_i \ldots X_s, m_1 \ldots m_n) \tag{1}$$

where X_i may be the gradient dX_{i-1}/dx. Since the body cannot be uniform, the equation of state can only apply to the portions of the body which are uniform. These would generally be planes of thickness dx (but could be lines or even points). The property of the body as a whole (for example, volume) would be equal to the summation of these incremental portions over the whole body.

The Separability of the Reversible and Irreversible Processes

Consider a body with a gradient and through which energy is constantly flowing. Then consider an incremental portion of that body which is uniform and hence for which the equation of state (1) is valid. Consider then an incremental change in the gradient imposed on the body. If this changes the properties of the incremental portion of body, this would result in a change in the internal energy equal to dE, and/or the absorption of some heat by the body equal to $-T\,dS$, and/or a change in volume with work done equal to $P\,dV$. If P and T are held constant during the application of this change in gradient there must have been reversible work done on the incremental portion of the body equal to a dG. This implies that the change in gradient must have some means for doing work on (or receiving work done by) the incremental body. It presumably does this by just utilizing some of the energy which is flowing through it (or adding energy to the energy flowing through the body) until a new steady state is reached.

Similar reversible paths for the body in a gradient can be constructed over other types of paths than isothermal, and the reversible components of the energy flows can be similarly identified. Hence the reversible and irreversible components of energy flows over defined thermodynamic paths are separable.

The Definability of the Thermodynamic Functions

The third requirement is that the thermodynamic functions such as A, H, G, and preferably GG be valid for sections of bodies in the presence of a gradient. For this to be true for the incremental bodies described above it is necessary that the components of these functions be real and valid. These components are E, P, V, T, S, X, and Y. The validity of the quantities P, T, and V is self evident. The quantity E is the sum of the kinetic and intermolecular potential energies per unit weight (plus the energy within the atoms). As such it has the same validity in the presence of a gradient as it does in "equilibrium" thermodynamics.

The existence of entropy S is less obvious. Some authors state that entropy cannot be defined when a body is not at equilibrium (Meixner, 1969). If the molecules of a body have a different average conformation in the presence of a gradient than they have without the gradient, then there exists the probability that the conformation will occur by chance and chance alone without the gradient. Hence, the statistical definition of entropy is valid for bodies in the presence of a gradient.

The validity of the thermal definition of entropy can also be established. In the above description of the reversible isothermal path accompanying a change in a gradient it was shown that a reversible absorption of heat was involved. This heat is equal to $-T\,dS$ which is equivalent to the conditions in "equilibrium" thermodynamics. The gradient drives the molecules from some average position about which the molecules move. The term $-T\,dS$ is the heat equivalent of the work the molecules will do to attain the average position that they have with no gradient.

The variables X and Y are less definite. The first must be a force variable associated with the gradient, and the second would be its conjugate extensive variable. A discussion of these will be delayed; however, it is possible to establish a limited thermodynamics without the extended thermodynamic functions GG etc.

The criterion of balance for the whole body in the presence of gradients is the summation of a particular work function (i.e., a free energy) for the whole body, which must be minimized. This is required since one part of the body is in balance with the other parts. It is analogous to the requirements for equilibrium of a multiple phase system in which the sum of the work functions of all the phases must be minimized. See Chapter VI. The particular work function (i.e., its free energy) which must be minimized depends on the system — i.e., which variables are held constant. This can be a rather complex problem.

Thermodynamic Balance in a Body with a Thermal Gradient

Consider a simple case of a body at constant pressure with a steady state thermal gradient. Such a body is shown in Figure 1. One end of the body is held at T_1 and the other is held at T_2. The body is uniform in cross section, uniform in chemical composition, and no heat escapes through the sides. Each cross section of such a body is at a uniform temperature and a uniform thermal gradient. The equation of state for that cross section is,

$$B = f(P, T, dT/dx) \tag{2}$$

When heat starts flowing through that cross section some of that heat can be diverted into a stored potential (just as with a spring) until a steady state is established. This potential energy is stored as a change in internal energy dE, as a change in the heat absorbed $-T\,dS$, and work done with a change in volume $P\,dV$. The sum is equal to dG, which is what should be expected for the reversible available energy at constant temperature and pressure. It is exactly analogous to the elastic energy of a stretched elastic body.

The gradient cannot be used to form an XY term to generate the extended thermodynamic functions, such as the GGibbs free energy function GG. This is because there is no reasonable term whose product with the temperature gradient has the dimensions of energy. However, we can presume there must be a function of the gradient $XX(dT/dx)$ with the dimensions of stress and a second function of the gradient $YY(dT/dx)$ with the dimensions of strain. The product of these two functions would then have the correct dimensions of energy and they could be used to define the GGibbs free energy function in a thermal gradient,

$$GG = E - TS + PV + (XX(dT/dx))(YY(dT/dx)) \tag{3}$$

A test for balance at constant P, T, and XX would be a minimum GG. If each individual cross sectional increment of the body met these requirements of constancy, then the whole body would be in balance or in the steady state. Such a constancy requirement would be that each of these sections would be at a different though constant temperature.

However, the constant T and XX are not realistic restrictions for bodies under a temperature gradient. The correct restriction is that the rate

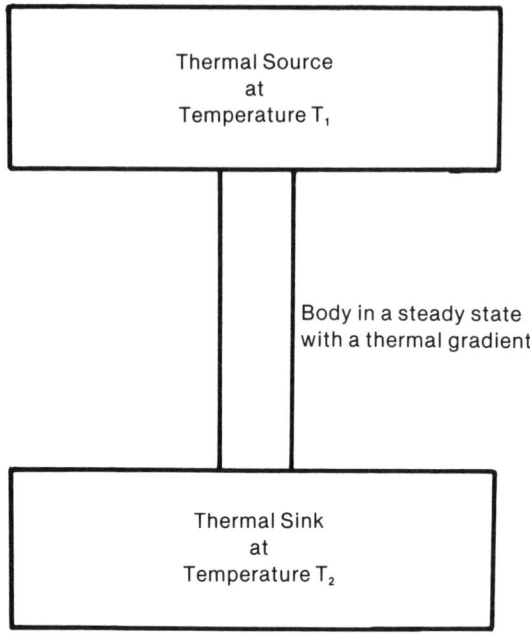

FIGURE 1. A body in a steady state in the presence of a gradient. Such a gradient may exist in a gas, or a liquid, or a solid, though conditions must be such that convection currents are not generated. Note that in this example all three components are held at constant pressure.

at which heat passes through (for a body of constant cross section) must be the same for each cross section, or

$$dq/dt = k(dT/dx) = \text{constant} \qquad (4)$$

must be identical for each cross section, where t is time and q is the amount of heat transferred through a section and k is the thermal conductivity. The thermal conductivity may vary with temperature and the temperature gradient, though dq/dt must remain constant throughout the body.

When a rubber body is stretched heat is normally released. An analogous release of heat may occur when a body is placed under a temperature or pressure gradient. Such a release of heat is observable when a gas diffuses through a porous plug or small capillary. A temperature change (usually a rise) occurs when the gas enters the capillaries and the opposite change occurs when the gas leaves. The amount of heat released is equal to the amount of heat absorbed, and it is called the heat of transport Q^*. Under the concept that the temperature gradient acts on the gas in a way analogous to a stress on a solid, Q^* would be equal to the $-T\,\Delta S$ component of the Gibbs free energy, ΔG^{PT}. If the imposition of the gradient is adiabatic rather than constant temperature the free energy would be ΔE^{SV} for constant volume or ΔH^{SP} for constant pressure (see Table I Chapter VIII).

There is a similar release and absorption of heat when a gas diffuses through a membrane in which it is slightly soluble. With a membrane this temperature change would involve a heat of solution; however, in the opinion of Denbigh (1951) other phenomena are involved. In the light of the above, one other phenomenon would be the change in free energy associated with the gradient in the membrane.

The Link between Temperature and Pressure Gradients

Consider two bodies of gas held at constant volume but at different temperatures. They are connected by a capillary or porous plug. See Figure 2. The diameter of the capillary or the pores in the plug should be much less than the mean free path of the gas molecules. The gas will generally flow from the body of gas at the lower temperature to the body of gas at the higher temperature until a small differential pressure is built up so that a steady state is reached at zero flow of gas.

Denbigh develops a relationship between the temperatures and pressures which is based on Thompson's hypothesis that the sum of the differential entropy terms at the steady state is equal to zero. He also states this derivation is questionable but that the relationship derived between the temperatures and the pressures has been verified by experiment.

I agree that his derivation is questionable and present an alternative derivation. It is based on two concepts, (a) the temperature gradient stores a potential energy within the body with the imposition of the gradient (analogous to a strain energy), and (b) the reversible processes can be separated from the irreversible and the reversible processes can be treated in terms of the steady state thermodynamics which have been proposed in the previous chapters.

Consider that in the model shown in Figure 2. A small amount of gas is transferred from vessel II to vessel I. This amount of gas has the volume dV. The amount of work done on the material in the capillaries on leaving vessel II is $(P + dP) dV$, while the work done on this same amount of gas entering I is $P dV$. The difference $dP\, dV$ is the work done on the gas in the capillaries.

It is reasonable to assume that all the work done on the gas in the capillaries is entropic work. This entropic work can be calculated as follows. The portion of a mole of gas transferred is dV/v so the heat transferred is $(Q^*/v)dV$. This represents a difference in the entropy at the inlet and the outlet of the capillaries of,

$$dS = (Q^*/(T + dT)v)dV - (Q^*/Tv)dV \\ = - Q^* dT\, dV/vT^2 \tag{5}$$

Converting the entropy change into its work equivalent by multiplying by T, we have,

$$dP\, dV = T\, dS = - TQ^*\, dT\, dV/vT \tag{6}$$

which converts to,

$$dP = -Q^*dT/vT \qquad (7)$$

which is the relation developed by Denbigh. He carries the development further by use of the relationship,†

$$Q^* = -\tfrac{1}{2}RT \qquad (8)^*$$

developed from kinetic theory to obtain the relationship,

$$P_1/P_2 = (T_1/T_2)^{1/2} \qquad (9)^*$$

which has been verified experimentally.

Equation (7) is stated to be in good agreement with the data on the fountain effect of Helium II. See Denbigh. The large value of Q^* would indicate in terms of the above theory that a temperature gradient in HeII causes a relatively large potential energy to be stored in HeII when placed under a thermal gradient. Such a potential energy would not necessarily be entropic, for a change in intermolecular potential energy (i.e., a dE) could be involved. The validity of equation (7) would, however, indicate that in this particular case the potential energy was primarily entropic.

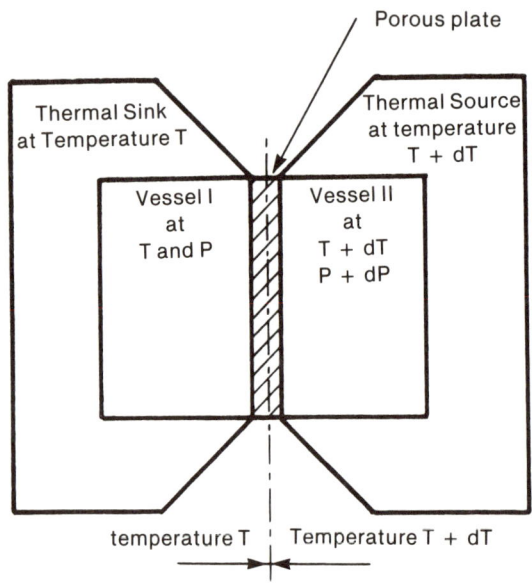

FIGURE 2. Two bodies consisting of the same gas are connected by a porous plate but are held at two different temperatures. Each body is held at constant volume. At the steady state a differential pressure builds up between the two bodies.

---

†*And assuming* $vP = RT$, i.e., that the gas is ideal.

It is worthwhile to speculate on the molecular source of the entropy changes in the above examples. In a temperature gradient the molecular kinetic energy should be greater in the down direction of the gradient than in the up direction of the gradient. This is the mechanism by which heat flows down the gradient. It is not necessary that any migration of the molecules exist for such a distribution of kinetic energy. However, in terms of a two independent variable equilibrium state, such a distribution of kinetic energy has a low probability. Would not the kinetic energy of the molecules do work in order to return the distribution of kinetic energy to a more probable distribution?

In a second example consider a monatomic gas in a capillary tube whose diameter is much less than the mean free path of the molecules. The molecules whose motion is across the tubes are likely to be deflected by the wall before their collision with other molecules. These molecules are more likely to have their direction changed than those whose velocity is parallel to the axis of the tube. Hence, the component of kinetic energy parallel to the axis of the tube is greater than the component of kinetic energy perpendicular to the axis of the tube. If the gas were outside the capillary this would be an improbable distribution of molecular velocities. Would not the kinetic energy of the molecules do work, in order to return the kinetic energy of the molecules to a more probable distribution?

The Soret Effect

If the body subjected to a temperature gradient (as in Figure 1) is a multicomponent liquid, a concentration gradient is developed. This occurs with salts in water as well as with polymers in their solvents. Such a development of a concentration gradient (which is a demixing process) clearly causes a decrease in entropy. In effect the temperature gradient does work against the entropy (i.e., kinetic energy) spring. This demixing is a reversible process just as mixing and demixing can be made reversible by the use of a semipermeable membrane. See the discussion on mixing in Chapter VIII.

If two components of a solution release heat on simple (i.e., irreversible) mixing this shows that the intermolecular forces have done work on the molecules, forcing them to a less random position. Any demixing will require that an equivalent amount of work be done against the intermolecular forces. If there is a change in volume, energy must be supplied or received from this action. If a thin section of the solution perpendicular to the direction of the gradient is visualized as being held at constant pressure and temperature, the Soret effect is associated with free energy changes which are very similar to those which occur with the straining of rubber. A spring model with a molecular kinetic energy spring and a molecular force spring in parallel is valid as a model for the mixing process. A dashpot is not needed in the model, as the viscous processes which dissipate work are very small if they exist at all.

The discussion of the Soret effect above has mentioned only the entropy changes associated with the development of concentration gradients. A second type of entropy change can occur simultaneously — one associated with molecular orientation. A temperature gradient can orient some types of molecules just as shear strain or magnetic or electric fields. The property of developing an orientation in a temperature gradient depends on the molecular structure. Internal energy changes, heat effects, and volume changes can all be associated with molecular orientation. At constant temperature and pressure the molecular orientation thus generates a Gibbs free energy. The molecular orientation has been reported only with some polymer solutions while the Soret effect occurs with all types of solutions.

Shear Stress Gradients in Viscoelastic Fluids

With viscoelastic fluids shear stress gradients can also be designated as velocity gradients, or as rate of shear gradients. These are essentially equivalent; however, it is preferred to use the name which designates the independent variable of state, and which is in turn designated in the equation of state. For this discussion the equation of state is selected to be,

$$B = f(P, T, X, dX/dx) \qquad (10)$$

so this discussion will be in terms of the shear stress gradient dX/dx.

Shear stress gradients in viscoelastic fluids generate gradients in the free energy associated with the elastic deformation of the fluid. If the conditions are constant temperature and pressure this is a gradient in the Gibbs free energy ΔG^{PT}. If the conditions are at constant S and P (a pseudo adiabatic condition) the gradient in free energy is a gradient in ΔH^{SP}. A gradient in free energy creates an unstable condition, for fluids tend to flow from the high free energy areas to areas of lower free energy. This must be one cause of the unstable flow which occurs during extrusion at high rates of shear. A stable flow of a one component viscoelastic fluid, in the presence of a shear gradient, must be regarded as a metastable state.

If a viscoelastic fluid has two or more components any free energy associated with elastic deformation is usually *not* distributed between the components in proportion to their mol fractions. For example, in the case of a polymer dissolved in a solvent such an (elastic) free energy will be primarily associated with the polymer component. The change in the conformation of the molecules in the presence of a shear stress gradient will be primarily in the polymer and little if any in the solvent. Hence a migration of the polymer molecules from areas of high shear stress to areas of lower shear stress will lower the total free energy of elastic deformation. This migration creates a concentration gradient.

If the material subject to a shear stress gradient is a molten polymer with a wide molecular weight distribution this will also result in the migration of the components with the highest free energy from areas of high shear stress to areas of lower shear stress. Schreiber, Storey and Bagley (1966) describe how they confirmed this effect with polyethylene. They extruded two types of polyethylene in a capillary viscometer — one polyethylene with a broad molecular weight distribution (a blend) and a second with a much narrower molecular weight distribution. The first showed a lowering of the molecular weight in the surface layer of the extrudate, and this increased with the length of the capillary. The second showed negligible effects. The first showed a large die swell which decreased rapidly with the length of the capillary, while the second showed much smaller effects. Aubert and Tirrell (1980) also discuss the relationship between shear stress gradients and concentration gradients and review some of the more recent literature on the subject.

Summary and Discussion

It has been shown that the action of a gradient on a body is analogous to the action of a stress. The gradient can generate a free energy, which if constant pressure and temperature are assumed, is equal to a ΔG^{PT}. However, (a) constant P and T are not realistic restraints (even for the thin sections of a body in a gradient), and (b) the gradients are not force-like variables which have associated conjugate variables with which the product has the dimension of work. Hence the thermodynamic criterion for balance in the steady state is more involved than the criterion for equilibrium of a body in three-variable thermodynamics.

The correct criterion of balance for a body in a temperature gradient must be, a minimum sum of free energies over the whole body in the gradient. The restraints on the system are: the heat flow to the body must equal the heat flow out, and the two extremes of temperature are fixed. An exact expression for this sum of free energies has not been developed.

It has been shown that gradients in one variable can generate gradients in other variables. This can be generalized by stating that gradients in all thermodynamic variables may be linked together — though any actual linkage would depend on the system. Thus we should expect that linkages could occur between gradients in: temperature, pressure, concentration, electrical potential, and stress.

It has been shown that temperature gradients generate concentration gradients. The generation of these concentration gradients decreases the entropy of the system. This change in the entropy of the system is one component of the free energy generated by the temperature gradient. This type of relationship should be expected as possibly occurring when any of the above mentioned gradients are linked together.

In all the examples discussed in this chapter it has been possible to separate the reversible action on bodies (as they are changed from one steady state to another) from the irreversible actions which occur at the same time. This is a further demonstration of the statement made in previous chapters that, although the reversible and irreversible processes may be linked, they are separable.

Chapter XI
The Relationship between the Thermodynamics of Rheology and Chemical Thermodynamics

There are three aspects to the relationship between the thermodynamics of rheology and chemical thermodynamics. First is their similarity, second is their contrasting difference, and third is their interaction when both elastic deformation and chemical reactions are involved.

The prime similarity is that the equations which describe the thermodynamics of rheology and chemical thermodynamics are almost identical. The prime difference is in the meaning of the mathematical equations in terms of physically measurable properties. They interact when both elastic deformation and chemical reactions occur at the same time.

The Equations of State

The equation of state for a body at equilibrium in terms of two independent variables of state is,

$$B = f(P, T) \tag{1}$$

Equilibrium with two independent variables permits the presence of no elastic deformation and requires that all chemical reactions have been taken to completion.

The equation of state for a uniform body at equilibrium under stress is

$$B = f(P, T, X_3, X_4) \tag{2}$$

where X_3 and X_4 are stresses normal to each other.

It is usually convenient to discuss the thermodynamics of rheology in terms of three independent variables of state when the equation of state is,

$$B = f(P, T, X_3) \tag{3}$$

where X_3 is a stress such as tension, compression or shear.

When chemical reactions are involved (instead of mechanical stresses) the most convenient form of the equation of state is,

$$B = f(P, T, \xi) \tag{4}$$

where ξ is the extent of the chemical reaction. This is an extensive variable. It is usually preferable to state the equation of state in terms of a force like variable which is a conjugate of the extensive variable of state. This conjugate variable is the affinity **A**, a thermodynamic variable used primarily by the French school (Prigogine and Defay, 1954). The affinity is,

$$\mathbf{A} = -(\partial G/\partial \xi)_{TP} = -\sum \nu_i \mu_i \qquad (5)$$

where ν_i is the stoichiometric coefficient of component i and μ_i is the chemical potential of component i. We can then write the equation of state as,

$$B = f(P, T, \mathbf{A}) \qquad (6)$$

Chemical reactions usually cannot be held at any arbitrary extent of reaction, though an elastic strain can be held at any arbitrary value of strain. Neither can most chemical reations be held at any arbitrary value of the affinity, though their analog stress can be held at any arbitrary value. The equations of state (4) and (6) apply to a particular instant of time while the chemical reactions is taking place. It is not a steady state. It is a state which in reality is usually reproducible from only one direction of the chemical reaction.

It is conceivable to have two chemical reactions which are independent of each other although they proceed at the same time in the same thermodynamic body. There are then four independent variables of state and the equation of state is then,

$$B = f(P, T, \mathbf{A}_3, \mathbf{A}_4) \qquad (7)$$

which is analogous to equation (2), the equation of state for stress.

The Free Energies and Reversible Work

The free energy of elastic deformation is produced by a stress X acting on an elastic body. The reversible work done on the thermodynamic body at constant temperature and pressure is (see Chapter VII — The Thermodynamics of Elastic Deformation),

$$\Delta G_e^{PT} = \Delta E^{PT} - T \Delta S^{PT} + P \Delta V^{PT} \qquad (8)$$

where the subscript e designates the ΔG_e^{PT} is the free energy of elastic deformation.

In elastic deformation the stress X varies from zero to some function of the strain Y. The work of elastic deformation at constant temperature and pressure is also equal to,

$$\Delta G_e^{PT} = -\int_o^1 X^{(PT)} dY = -\bar{X}^{(PT)} \qquad (9)$$

where $\bar{X}^{(PT)}$ is the average force exerted over a strain of one unit at constant temperature and pressure.

Reversible paths do not exist for most chemical reactions. Chemical reactions usually proceed over irreversible paths during which no external work is done (except that portion associated with a change in volume). These irreversible paths of chemical reactions are analogous to those in rheology as for example (a) The complete release of a stress such that it allows a body to return to an unstrained state without doing external work, or (b) Molecular relaxation in which a stress is lowered by molecular rearrangements through which the stress is released without changing the strain or doing external work.

With chemical reactions it is convenient to consider that a force like variable **A**, the affinity, can do reversible chemical work on a thermodynamic body, displacing a chemical reaction in an amount $-\xi$. The reversible chemical work is then equal to,

$$\Delta G_c^{PT} = \Delta E^{PT} - T \Delta S^{PT} + P \Delta V^{PT} \qquad (10)$$

where the subscript c designates that the free energy is that of a chemical reaction.

The reversible chemical work is also equal to

$$\Delta G_c^{PT} = -\int_0^1 \mathbf{A} \, d\xi = -\bar{\mathbf{A}} \qquad (11)$$

where $\bar{\mathbf{A}}$ is the average affinity over that unit path. This corresponds to equation (9).

With a viscoelastic fluid there are also no real paths by which all the stored elastic energy is recoverable as external work. The rate of molecular relaxation is too great to recover that stored elastic energy. One must use imaginary paths, in which molecular relaxation is temporarily frozen, over which the stored elastic energy is recoverable. Thus imaginary reversible paths are useful for the discussion and explanation of both the chemical and rheological free energies.

Other Thermodynamic Equations

In the above discussion it has been shown that the affinity **A** is analogous to the stress X and the extent of a chemical reaction ξ is analogous to the strain Y. This analogy is complete in that it is true for practically all thermodynamic expressions. For example, many expressions in this book can be converted to the expressions in Prigogine and Defay by this substitution. The following illustrates some of these conversions with references to the equivalent equations in this book and Prigogine and Defay (noted as P&D).

P&D develop their equations somewhat differently than they are developed here; however, the results are the same. They do give one equation of state in which the extent of reaction is included as a variable of state,

$$E = f(T, V, \xi) \qquad (12)$$

(P&D equation 2.5, note that the symbols used have been changed to those of this book.) P&D are careful to note the two variables which are held constant in the partial differentials as is done in this book.

From the first law,

$$dE = T \, dS - P \, dV - \mathbf{A} \, d\xi \qquad (13)$$

(refer to P&D 4.23 and VII-35 in this book)

The total differential of G is,

$$dG = dE - T \, dS - S \, dT + P \, dV + V \, dP \qquad (14)$$

Subtracting (13) from (14) we obtain

$$dG = -S \, dT + V \, dP - \mathbf{A} \, d\xi \qquad (15)$$

(refer to P&D 4.26, and VII-38 in this book)

Similarly we obtain,

$$dH = T\,dS + V\,dP - \mathbf{A}\,d\xi \tag{16}$$

$$dA = -S\,dT - P\,dV - \mathbf{A}\,d\xi \tag{17}$$

(P&D equations 4.24 and 4.25, and VII-36 and 37)

From these four equations we obtain the relationships

$$\mathbf{A} = -(\partial E/\partial \xi)_{SV} = -(\partial H/\partial \xi)_{SP} = -(\partial A/\partial \xi)_{TV} = -(\partial G/\partial \xi)_{PT} \tag{18}$$

(P&D 4.30 and VII-43)

We can also write four extended thermodynamic functions in which $\mathbf{A}$ replaces X and ξ replaces Y (see equations VII-20, 21, 22, 23).

$$GG = E - TS + PV + \mathbf{A}\xi \tag{19}$$

$$HH = E + PV + \mathbf{A}\xi \tag{20}$$

$$HA = E - TS + \mathbf{A}\xi \tag{21}$$

$$AA = E + \mathbf{A}\xi \tag{22}$$

From these extended functions and manipulations similar to the above and those given in Chapter VII we can obtain a set of twenty four Maxwell relations analogous to VII 49-72. Twelve of these Maxwell relations are given by P&D 4.36-4.39 in terms of the affinity and the extent of chemical reaction. P&D miss twelve of the Maxwell relations because they do not use the extended thermodynamic functions which are needed to develop the second twelve.

P&D proves to be a source of other equations which can be converted for application to rheology by merely substituting the stress X for the affinity and the strain Y for the extent of chemical reaction. For example, their equation 4.8 converts to,

$$(\partial(X/T)/\partial T)_{PY} = (1/T^2)\,(\partial H/\partial Y)_{PT} \tag{23}$$

A check of their development of 4.8 confirms that it also applies to elastic deformation. This equation could be useful for the study of the enthalpy changes, which occur on the stretching of polymers. Some polymers form crystallites on stretching and the effect of this would be reflected in the above equation (23).

Galvanic Cells and Chemical Reactions

Up to this point it has been assumed that chemical reactions are reversible only in an hypothetical way but not in reality. There are, however, chemical reactions which *are* reversible. One type of reversible chemical reaction is that which occurs in galvanic cells.

A galvanic cell usually contains three bodies — the cathode, the anode and the electrolyte. These bodies form a system and the extensive properties of this system are the sums of the extensive properties of its bodies. For example the free energy of all the bodies in the system, i.e., $\Sigma\,\Delta G_c^{PT}$.

In terms of two independent variables of state, equilibrium in a galvanic cell is attained only when the cell is completely discharged. If electrical work is done to recharge the cell at constant temperature and pressure, this will result in a change in the internal energy $\Sigma \Delta E^{PT}$. If this recharging of the cell results in a reversible exchange of heat with the surroundings this sums to $\Sigma - T \Delta S^{PT}$. If a change in total volume occurs, this is equal to $\Sigma P \Delta V^{PT}$. Thus the work done in recharging a galvanic cell is completely analogous to the work done in stretching an elastic body, and, (omitting the summation signs) we have,

$$\Delta G_c^{PT} = \Delta E^{PT} - T \Delta S^{PT} + P \Delta V^{PT} \tag{24}$$

The electrical work done in recharging the galvanic cell is also equal to

$$\Delta G_c^{PT} = -\int_o^1 \mathcal{F} \, nj \, \mathcal{E} \, dQ = - \mathcal{F} \, nj \, \overline{\mathcal{E}} \tag{25}$$

where $\mathcal{E}$ is the electromotive force, n is the number of moles, j is the equivalents per mole and Q is the quantity of electricity, $\overline{\mathcal{E}}$ is the average electromotive force, over one unit of electricity Q, and $\mathcal{F}$ is the Faraday constant.

The relationship between the free energy and the electromotive force of a galvanic cell is usually stated as,

$$\Delta G_c^{PT} = - nj \, \mathcal{F}\mathcal{E} \tag{26}$$

instead of the more precise,

$$(\partial \Delta G_c^{PT}/\partial Q)_{PT} = -nj \, \mathcal{F} \, \mathcal{E} \tag{27}$$

Equation (26) is valid for practically all galvanic cells for (if other factors do not change) their voltage remains constant as the reaction proceeds.† In other words the voltage does not depend on the quantity of the material in the electrodes.

Examination of equations (11) and (25) shows that these equations are equivalent. In other words the affinity **A** is equal to the voltage of a hypothetical galvanic cell in which chemical reactions are reversible (corrected for the constants in the equation).

The Free Energy and the Restraints on the System

Equation (18) states that the driving force of a chemical reaction **A** is the same for the four sets of restraints given by the equation. Since the voltage of a galvanic cell is equal to the affinity **A** this means, for example that the voltage of a cell held at constant temperature and pressure is the same as when it is isolated from its environment, i.e., held at constant entropy and constant volume.

Equation (VII-43) states the same for the retractive force of an elastically strained body. The force X is the same whether the body is held at

---

†Voltage is independent of the extent of the reaction in the cell if no concentration gradients, or change in concentrations or similar actions occur.

constant temperature and pressure, or constant temperature and volume, or at constant entropy and volume, or at constant entropy and pressure.

The fact that the driving force is the same, however, does not mean that the total available work is likewise independent of the restraints on the system. Their non-equality can readily be shown with an example. Consider a rubberlike elastic body held at constant temperature and constant volume. This body is strained, and then allowed to recover under the same conditions — to obtain the maximum work. In Figure 1 the solid line shows a hypothetical plot of the retractive force against strain. The maximum work done is the area under the curve. If however the body is allowed to recover over a different path — constant entropy and constant volume — the temperature will be lowered as the work proceeds. Since the force X is proportional to the absolute temperature at any particular strain Y, the force on the second path will be lower than the first at all values of strain between the two end points on the paths. The second path is shown as the dotted line in Figure 1. Hence, the maximum available work over the second path is less than over the first path.

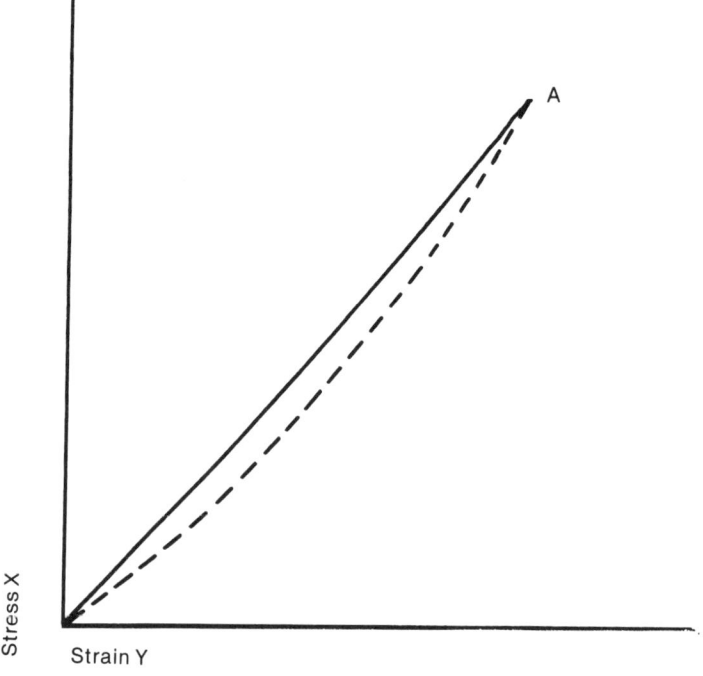

FIGURE 1. An hypothetical plot of stress versus strain for (a) an isothermal path — the solid line and (b) an adiabathic path — the dotted line. If the initial states are identical at point A, the maximum available work for the isothermal path is always greater than that of the adiabatic path. The maximum available work is the area under the curve. This results from the lower temperature on the adiabatic path, and from the proportionality of stress to absolute temperature for rubberlike materials.

The maximum available work at constant temperature and pressure is,

$$\Delta G^{PT} = \Delta E^{PT} - T\Delta S^{PT} + P\Delta V^{PT} \tag{28}$$

at constant temperature and volume,

$$\Delta A^{VT} = \Delta E^{VT} - T\Delta S^{VT} \tag{29}$$

at constant entropy and pressure

$$\Delta H^{SP} = \Delta E^{SP} + P\Delta V^{SP} \tag{30}$$

and at constant entropy and volume

$$\Delta E^{SV} \tag{31}$$

Each of the above four expressions is the maximum available work from the same body under different restraints, i.e., over different paths. They are free energies. It has already been shown by Figure 1 that (29) and (31) are not equal. It can be shown that none is equal to the other. This should be expected since with the same initial condition of stress X the path of two independent variable equilibrium over the various paths will lead to non-equivalent states. Hence,

$$\Delta G^{PT} \neq \Delta A^{TV} \neq \Delta H^{SP} \neq \Delta E^{SV} \tag{32}$$

Equation (32) is valid for both: — (a) when the maximum available work is associated with elastic strain and (b) when the maximum available work is associated with a chemical reaction.†

The Interaction of the Chemical and Rheological Free Energies

When both a chemical reaction and an elastic strain are involved the equation of state is,

$$B = f(P, T, X, \mathbf{A}) \tag{33}$$

Two of the variables represented by B are the free energy associated with a chemical reaction, and the free energy associated with an elastic strain (a rheological free energy). The actual maximum available work (i.e., the free energy) depends on the restraints on the system.

Consider a thermodynamic body on which a weight is hanging and which is kept at constant pressure and constant temperature. The weight exerts a constant stress X on that body. In this body it is hypothetically possible to use some outside agent to reverse a chemical reaction in the body. This reversal would be analogous to the charging of a galvanic cell. The work done in reversing the chemical reaction would equal a free energy, and is equal to

$$\int \mathbf{A} \, d\xi \tag{34}$$

---

†The inequality of the various free energies does not imply there is any question of validity for the many specific uses of the Gibbs free energy ΔG^{PT}.

Some of the energy used in reversing the chemical reaction would increase the internal energy in an amount ΔE^{PTX}; heat may also be absorbed in the process in an amount of $-T \Delta S^{PTX}$; if there is a change in volume work will be done in the amount $P \Delta V^{PTX}$; and if the changes cause the weight to be raised or lowered work will be done in the amount of $X \Delta Y^{PTX}$. These equal equation (30) but also sum to,

$$\Delta GG_c^{PTX} = \Delta E^{PTX} - T \Delta S^{PTX} + P \Delta V^{PTX} + X \Delta Y^{PTX} \qquad (35)$$

If the strain Y had been held constant instead of the stress X, the expression for the free energy would be,

$$\Delta G_c^{PTY} = \Delta E^{PTY} - T \Delta S^{PTY} + P \Delta V^{PTY} \qquad (36)$$

In the terminology of this paper these are the GGibbs and the Gibbs free energies of the chemical reaction.

In a similar way it can be shown that the GGibbs free energy associated with the elastic strain at constant affinity $\mathbf{A}$ is

$$\Delta GG_e^{PT\mathbf{A}} = \Delta E^{PT\mathbf{A}} - T \Delta S^{PT\mathbf{A}} + P \Delta V^{PT\mathbf{A}} + \mathbf{A} \Delta \xi^{PT\mathbf{A}} \qquad (37)$$

and the Gibbs free energy associated with elastic strain but at constant extent of reaction ξ is,

$$\Delta G_e^{PT\xi} = \Delta E^{PT\xi} - T \Delta S^{PT\xi} + P \Delta V^{PT\xi} \qquad (38)$$

With the change of the Δ to a d in equation (35) we can obtain,

$$(\partial GG_c^{PTX}/\partial Y)_{ESV} = X \qquad (39)$$

and from equation (37) we can write,

$$(\partial GG_e^{PT\mathbf{A}}/\partial \xi)_{ESV} = \mathbf{A} \qquad (40)$$

Other relationships can be found by standard methods. We can define the thermodynamic functions,

$$GGG = E - TS + PV + XY + \mathbf{A}\xi \qquad (41)$$

$$GG_c = E - TS + PV + XY \qquad (42)$$

$$GG_e = E - TS + PV + \mathbf{A}\xi \qquad (43)$$

From the first law

$$dE = T\,dS - P\,dV - X\,dY - \mathbf{A}\,d\xi \qquad (44)$$

If we take the complete differential of (42) and then subtract (44) from it we obtain,

$$dGG_c = -S\,dT + V\,dP + Y\,dX - \mathbf{A}\,d\xi \qquad (45)$$

From this we obtain,

$$(\partial GG_c/\partial \xi)_{TPX} = -\mathbf{A} \qquad (46)$$

In the same way we obtain from (39),

$$(\partial GG_e/\partial Y)_{PT\mathbf{A}} = -X \qquad (47)$$

A test for balance (the equivalent of equilibrium) with four independent variables of state is that,

$$\Delta GGG^{PTX\mathbf{A}} \text{ is at a minimum} \tag{48}$$

This would not be equilibrium with the usual terminology for it is usually impossible to hold a chemical reaction at constant affinity **A** (or likewise at constant extent of reaction ξ). Chemical reactions usually proceed at a rate governed by the kinetics of the reaction.

The Mechanicochemical System

The mechanicochemical systems (Katchalsky, 1965) convert chemical energy directly into work, or conversely convert mechanical work into chemical energy. Muscles are an example of such systems. Fibers and gels whose swelling is controlled by chemical agents are other examples.

The equation of state (33) in which both X and **A** are variables of state gives one basis for these systems. This equation implies that when a body is strained it changes the equilibrium point of chemical reactions in that body and vice versa.

The mechanicochemical cycles can be diagramed in the same way that Carnot cycles can. Figure 2 (a), (b), and (c) shows such a diagram. Point (1) is a position with high elatic free energy (ΔG_e^{PT}) and high chemical free energy (ΔG_c^{PTY}). The body is allowed to do mechanical work keeping the extent of the chemical reaction constant. It goes to point (2) on the diagrams. The chemical reaction is then allowed to proceed keeping the strain Y constant during which it goes to point (3). Mechanical work is then done on the body keeping the extent of chemical reaction constant and returning the strain to its initial value. The body is then at point (4). The cycle is then completed by reversing the chemical reaction at constant strain and returning to point (1).

When all paths are completely reversible the area in (b) is equal to the mechanical work done by the body, and the area in (c) is equal to the chemical work done on the body. With reversible paths these two areas are equal when expressed in the same units of energy.

Reversing the extent of a chemical reaction as is required in these cycles requires a source of energy. This could conceivably come from electrical work done on the body. It could also come from chemical sources. For example, changing the pH of the solution in which a gel or a fiber is immersed could modify the extent of a chemical reaction within the body. Such actions are known to cause such bodies to swell or shrink.

In the cycle described by Figure 2, either the strain or the extent of chemical reaction is held constant on the paths shown. They are also at constant P and T. The free energies are thus Gibbs free energies, either ΔG_c^{PTY} or ΔG_e^{PT}. It is also possible to construct a cycle in which either the stress X or the Affinity **A** is held constant. In this case the free energies involved would be ΔGG_c^{PTX} and $\Delta GG_e^{PT\mathbf{A}}$. Cycles in which V is held constant instead of P, and in which S is held constant instead of T, can also be constructed.

Since temperature changes are not involved in these cycles they have a theoretical efficiency of one hundred percent.

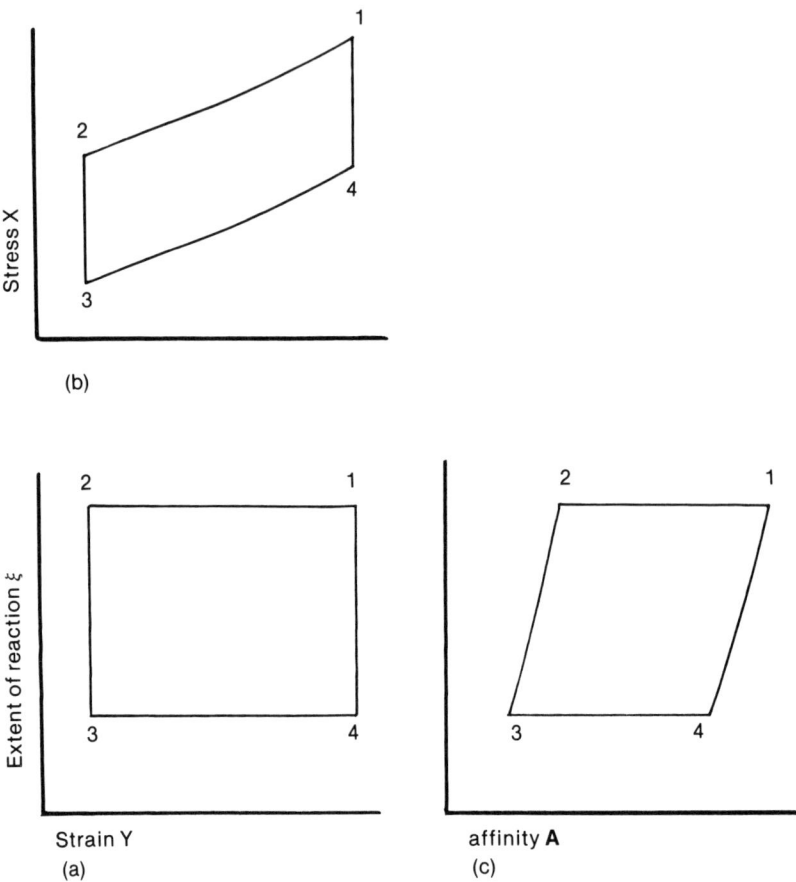

FIGURE 2. A diagram of a mechanicochemical cycle. (a) shows the cycle in terms of extent of chemical reaction versus strain, (b) shows the mechanical cycle, and (c) shows the chemical cycle.

Chapter XII
Recapitulations and Speculations

My first encounter with the problems of rheology was simple in statement but complex in solution — how to measure the properties of an industrial material which control its behavior in flow. There are a number of possible approaches to such a problem but the one which intrigued me the most was the thermodynamic approach. The viewpoints which finally developed have been given in the preceding chapters. They are conservative viewpoints developed from: the principles of mechanics, the first and second laws of thermodynamics, and the mathematical concepts or probability. These principles are applied to models of systems containing bodies which follow the observed behavior of real substances. The behavior of these bodies is then interpreted in a qualitative way in terms of molecules which follow the laws of Newtonian mechanics.

On the other hand the viewpoints developed here are different than the consensus. Because of this, care is taken to state the starting assumptions, and then build conceptions and conclusions step by step. Whenever possible, concepts are examined from several viewpoints. This both confirms the validity of the concepts and helps the reader form a better understanding of them.

Some Concepts and Viewpoints Reviewed

In this book a number of special concepts or viewpoints are emphasized. They may appear to be elementary and none of them are particularly new. However, the conclusions resulting from their combination are new. An outline of these special concepts follow.

(a) A thermodynamic model is set up (or hypothesized) which is closely analogous to a real system for which thermodynamic equations are desired. These models are similar to the free bodies of mechanics.

(b) All the variables of the model are classified into,
  (1) Force-like intensive variables.
  (2) Conjugates of the force-like variables. These are extensive variables and the product of the two conjugate variables i.e., (1) and (2), is energy.
  (3) Internal energy and other variables which are not paired with a conjugate variable.
  (4) Gradients.

(c) The equation of state of the thermodynamic body in the system is obtained from the model and the list of variables. One independent variable of state is included in the equation of state for each pair of conjugate variables in list (b) above.

(d) When developing a thermodynamic equation for a system it is usually convenient to hold one of each of the two conjugate variables constant. This applies in particular to the Δ or difference functions. The variables which are held constant are the restraints on the system.

(e) The maximum work available from a body (i.e., its free energy) depends on the restraints which have been assumed for the system in which it is placed. (The concept of multiple free energies does not of course invalidate the many uses of the Gibbs free energy (ΔG^{PT}) such as the calculation of the equilibrium constants, etc.)

(f) The state of equilibrium (or balance in the steady state) depends on the restraints which have been assumed for the system, since equilibrium is that state with the minimum of a particular free energy.

(g) The various thermodynamic functions (conventional and extended) are useful for defining the various free energies. They provide the most convenient means for defining the free energies and that is one of their main purposes.

(h) The equations of prime interest are those which describe the effect of the actions of the system on the *properties* of the body or bodies in that system.

(i) The properties of a body in the steady state can be treated in principle the same as the properties of a body in equilibrium. The difference is that large bodies in the steady state are not uniform so that the equations of state apply to portions of these bodies which are uniform (thin sections etc.) or apply to asymtotic limits of bodies as their thickness approaches zero.

(j) The difference between two uniform bodies in steady state is constant and describable in terms of reversible actions along hypothetical steady state paths between the two steady states. Thus models are developed in which the reversible and irreversible processes are separable.

(k) It helps a great deal to have a notation which denotes which variables are held constant for the difference equations (i.e., the Δ equations) as well as for the partial differentials. Such a notation aids greatly in the manipulation and interpretation of the meaning of the thermodynamic equations.

(l) It is important to use symbols and notations which are understandable and easily remembered. Although it has been necessary to use some nonstandard symbols and definitions, an effort has been made to keep differences in notation to a minimum. The use of the double letters for the extended functions has a logical relationship to the symbols used for the common functions. The addition of the superscripts to the delta quantities should add to the clarity of the expressions.

RECAPITULATIONS AND SPECULATIONS   141

(m) Frequent use is made of parallelisms and contrasts. Examples are: the parallelism between affinity **A** and the stress X, the parallelism between viscoelastic deformation and the models of the dashpot with two springs (Figures VII-10 and IX-3). A contrast is formed by the deficiencies of the spring and dashpot models.

Comments on Entropy and Chapter IV

The chapter on entropy (Chapter IV) is somewhat of a digression. This was originally an exercise by the writer to develop a better comprehension of entropy. It has done this for the writer and it is hoped it will do the same for others.

Equation IV-18 fits the data from the thermodynamic tables fairly well. However, this fit should not be considered an absolute proof of its validity; for, other equations which are quite different also fit the data nearly as well (Hull, 1973). Also the development includes an approximation. An understanding of the validity of this approximation can be gained by examining the relationships between the arithemetic mean molecular velocity and the geometric mean molecular velocity for a log normal distribution of molecular velocities. See equation III-6*. It is well known that for such a distribution the arithemetic mean velocity $\bar{v}$ is,

$$\bar{v} = (8kT/\pi m)^{1/2} \qquad (1)^*$$

By methods analogous to the way equation (1)* is obtained it can be shown that the geometric mean molecular velocity v' for a log normal distribution is,

$$\bar{v}' = \exp \left(\int_0^\infty 4\pi(m/\pi kT)^{3/2} v^2 \ln \bar{v} \, (\exp(-mv^2/2kT)) \, d\bar{v} \right.$$
$$= \exp((2 + \ln(kT/2m) - c)/2)$$
$$= (4.149 kT/m)^{1/2} \qquad (2)^*$$

where c is Euler's constant (0.577215).

Consider an ideal gas at T_1 and a second ideal gas at T_2. Note that the distribution of molecular velocities of an ideal gas is independent of volume or pressure. From equations (1)* and (2)*,

$$(\bar{v}_1/\bar{v}_2) = (T_1/T_2)^{1/2} = (\bar{v}_1'/\bar{v}_2') \qquad (3)^*$$

That is the arithemetic mean and the geometric mean are directly proportional to each other for a log normal distribution. Although the log normal distribution has been shown by experiment to be correct for dilute real gases, it is expected that the velocity distribution will vary from the log normal as intermolecular forces and molecular size affect the properties of heavy monatomic gases. Hence equation (3) must be considered an approximation over the constant energy paths used for the calculations on argon and xenon in Chapter IV.

Speculation on Entropy Changes in a Capillary

In Chapter X the thermodynamic behavior of a gas in a capillary was discussed. It was shown that if the mean free path of the molecules was much greater than the diameter of the capillary, the distribution of molecular velocities along the length of the capillary must be different than across the capillary; i.e., the distribution of velocities is not symmetrical.

One can speculate that such an unsymmetrical distribution could be reached by chance and chance alone, and that the probability that it could be reached by chance is the basis for calculating a change in entropy. Also, the molecules would use their kinetic energy to do work to return the velocity distribution to a symmetrical one.

There are other examples where velocity distributions are not symmetrical — surfaces and shock fronts for example. It may be worthwhile to develop the concept of the relationship of entropy to the symmetry of molecular velocities further.

It is interesting to speculate as to whether entropy has direction. It is certainly not a vector but at the same time entropy changes are associated with directions — for example, with molecular orientation in the flow of viscoelastic bodies as well as possibly in the distribution of molecular velocities.

Temperature as a Rheological Variable

The rheologist customarily makes his measurements at constant temperature. The thermodynamics of rheology suggests that he should pay attention to temperature as an inherent rheological variable, and hence treat it in greater depth than is customary. Both the adiabatic temperature changes, as shown in Figures VII-8 and IX-13, and the effect of temperature on the stresses (as done by Garner, Nissan, and Wood (1950) and discussed in Chapter IX), should become common rheological methods; and the results should be interpreted in terms of thermodynamics. These temperature effects should characterize the rheological behavior of polymers just as much as the stress-strain (or stress rate-of-change-in-strain) measurements do now. Of course it will be necessary to improve on the relatively crude tests for the temperature effects which have been described here.

Some Final Comments

The material in this book should not be considered complete. There are other thermodynamic relationships, other steady states, and other thermodynamic variables which have not been mentioned. The concepts on the relationship of entropy of heavy monatomic gases need to be extended to more complex molecules — to start with perhaps diatomic molecules.

The reader who expects to apply the thermodynamics rheology should not expect to find all the formulae here which will be useful to him. He can develop others by the methods demonstrated here, and he should also look for relationships which are analogous to those published for the thermodynamics of two independent variables of state. It is not possible to predict how the thermodynamics of rheology will develop. Its future path will depend on the needs and ingenuity of, the rheologists, the physicists, the chemists, and the engineers, who work in the area.

It would be gratifying if all the concepts in this book were quickly accepted by the scientific community; however, this will probably not occur. I only hope there are a sufficient number of readers who will follow the arguments so that the book will become known and discussed.

If there are errors here (as I am sure there must be) I would appreciate their being pointed out. I would also appreciate being informed of any points which readers feel are obscure, or which have not been adequately explained or established.

Appendix I
List of Principal Symbols

Extensive thermodynamic quantities are printed in large capitals (such as A, B, G). They are indicated by a # placed before them in the following list. When these extensive quantities are converted to a per unit weight basis they are printed as small capitals (such as A, B, G). When the molecular weight is known the unit weight basis is the molecular weight.

$A = E - TS$. The helmholtz free energy function.

$AA = E + YY$. An extended thermodynamic function.

B. A dependent variable of state, it may be either intensive or extensive.

$Cp = (\partial H/\partial T)_P$. The specific heat at constant pressure.

$Cpx = (\partial HH/\partial T)_{PX}$. The specific heat at constant pressure and constant stress.

$Cv = (\partial E/\partial T)_V$. The specific heat at constant volume.

$Cvy = (\partial E/\partial T)_{VY}$. The specific heat at constant volume and constant strain.

E. The internal energy.

E_k. The molecular kinetic energy.

E_p. The intermolecular potential energy associated with the intermolecular forces.

$G = E - TS + PV$. The Gibbs free energy function.

ΔG_c^{PT}. The Gibbs free energy of a chemical reaction.

ΔG_e^{PT}. The Gibbs free energy of elastic deformation.

G_o. Reference state for Gibbs free energy function.

$GG = E - TS + PV + XY$. Extended Gibbs free energy function. Also the GGibbs free energy function.

$H = E + PV$. The enthalpy.

$HA = E - TS + XY$. An extended thermodynamic function.

$HH = E + PV + XY$. The extended enthalpy.

K. The equilibrium constant or degrees Kelvin.

M. The molecular weight.

N. Number of molecules or number of points.

P. Pressure.

$Pi = (\partial E/\partial V)_T = (T(\partial P/\partial T)_V - P$. The internal pressure.

Q. Heat absorbed.

Q^*. Molar heat of transport.

R. Gas constant, or radius, or ratio.

\# S. Entropy.

ΔS. Change in entropy per mole.

$\Delta S''$. Change in entropy per mole calculated from equation IV-21.

T. Temperature on the absolute scale.

\# V. Volume.

W. Work in Chapter III, Probability of a combination of events.

X. The retractive force in elastic deformation†.

$Xi = (\partial E/\partial Y)_{TV} = T(\partial X/\partial T)_{YV} - X$. Internal stress in tension or in shear. It is directly analogous to internal pressure Pi.

\# Y. The conjugate of variable X. Strain.

A. Affinity (Prigogine and Defay, 1954).

e_{ki}. The kinetic energy of molecule i.

f. Function of. Force between molecules. Fugacity.

k. Boltzmann's constant, or thermal conductivity.

ℓ. Length.

m. Mass of one molecule.

m_n. The weight of component n.

n. Number of components as in the phase rule.

n_i. Constant in Gosman's equation of state for argon.

p. The number of phases in the phase rule.

q. Heat absorbed.

r. Degrees of freedom in phase rule, or radius in a capillary.

r. Average intermolecular distance.

s. Number of independent variables of state as used in the phase rule.

t. Time.

$\bar{v}$. Arithmetic mean of the molecular velocities.

$\bar{v}'$. Geometric mean of the molecular velocities.

w. Probability of a single event.

w_c. Conditional probability of a single event.

x. A distance.

$\dot{x}$. Velocity.

$\ddot{x}$. Acceleration.

x_1, x_2, x_3. The three orthogonal axes.

x_i. The mol fraction of component i.

α. Phase alpha.

β. Phase beta.

Δ. Difference or difference function.

θ. The thermodynamic temperature.

μ_i. The chemical potential of component i.

ν_i. The stoichiometric coefficient of component i in a chemical reaction.

ξ. The extent of a chemical reaction.

Π. The osmotic pressure.

ρ. The density.

Σ. Summation.

τ. Time.

ϕ. Angle.

$< >$. Indicates an expected value or a time average.

*. Equations marked with this sign have an ideal assumption — such as an ideal gas, an ideal solution, or an ideal rubber.

**. Equations marked with this double sign are specifically for the model described for a heavy monatomic gas.

$\mathcal{E}$. Voltage (of a galvanic cell).

$\mathcal{F}$. Faraday unit of electricity.

---

† This is usually referred to as the stress in this book, however, it is more precisely the negative of the stress.

References

All of these references were not quoted specifically in the text, however, they were used for the preparation of the book. The titles of papers in periodical publications are given so that they can be used for general references.

Adams, James, "Phase Transitions and Liquid Crystals", in *Phase Transitions 1973*, Proceedings of the Conference on Phase Transitions and Their Applications in Material Science, Pergamon Press, N.Y. 1973.

Aubert, James H., Tirrell, Matthew, "Macromolecules in Nonhomogeneous Velocity Gradient Fields", *J. Chem. Phys* **72**(4), pp. 2694-2701, 15 Feb. (1980).

Bagley, E.B., Birks, A.M., "Flow of Polyethylene into a Capillary", *J. Appl. Phys.*, **31**, 556-61, 1960.

Bagley, E.G., "The Separation of Elastic and Viscous Effects in Polymer Flow", *Trans. Soc. Rheol.* **5**, 355-68, 1961.

Bagley, E.B., Schreiber, H.P., "Effect of the Entry Geometry on Polymer Melt Fracture and Extrusion Distortion", *Trans. Soc. Rheol.*, **5**, 341-53, 1961.

Bagley, E.B., Birks, A.M., "Flow of Polyethylene into a Capillary", *J. Appl. Phys.*, **31**, 556-61, 1956.

Bergen, J.T. (ed), *Viscolasticity, Phenomenological Aspects,* Academic Press, N.Y. 1960.

Bird, R.B., Hassager, O., Armstrong, R.C., Curtis, C.F., *Dynamics of Polymeric Liquids, Volume II, Kinetic Theory,* Wiley, N.Y. 1977.

Bland, D.R., *The Theory of Linear Viscoelasticity,* Pergamon Press, N.Y. 1960.

Boltzman, Ludwig, *Lectures on Gas Theory,* translated by Stephen G. Brush, U. of Calif. Press, Berkeley 1964.

Bordoni, P.G., "On the Exact Relations between the Specific Heats of an Elastic Solid", *J. Rat. Mech. and Analy.*, **4**, 975-81, 1955.

Bridgman, P.W., *The Nature of Thermodynamics,* Harper, 1961.

Bridgman, P.W., *The Way Things Are,* Viking, 1961.

Bridgman, P.W., *The Thermodynamics of Electrical Phenomena in Metals,* Dover, 1961.

Brush, Stephen G., *The Kind of Motion We Call Heat,* Two volumes, North Holland, N.Y. 1976.

Bryand, G.M., Wakenham, H., "Entropy Changes Accompanying the Stretching of Cellulose Fibers in Water", *Text. Res. J.*, **25**, pp. 224-35, 1955.

Brydson, J., *Flow Properties of Polymer Melts,* Van Nostrand, N.Y. 1970.

Buchdahl, H.A., *The Concepts of Classical Thermodynamics,* Cambridge, 1966.

Callen, H.B., *Thermodynamics,* Chapter 13, "Solid Systems — Elasticity", Wiley, N.Y. 1960.

Casal, A., Porter, R.S., *Polymer Stress Reactions,* Two volumes, Academic Press, N.Y., 1979.

Chandrasekhar, S., "Stochastic Problems in Physics and Astronomy", *Rev. Mod. Phys.*, **15**, 1-89, 1943.

Cox, R.G., Mason, S.G., "Suspended Particles in Fluid Flow Through Tubes", *Ann. Rev. Fluid Mech.*, **3**, 291-316, 1971.

De Groot, S.T., Mazur, P., *Non-Equilibrium Thermodynamics,* Interscience, N.Y. 1962.

Denbigh, K.G., *Thermodynamics of the Steady State,* Wiley, N.Y. 1951.

Drigbaum, Dawkins, Via, and Balta, "Effect of Strain on the Thermodynamic Melting Temperature of Polymers", *Rub, Chem, and Tech.*, **40**, 788-800, 1967.

Dunlap, R.E., Pokigo, F.J., Glick, S.E., "Annealing Injection-Molded Styrene", *Modern Plastics*, Aug. 28, 1950.

Ehrenfest, Paul and Tatiana, *The Conceptual Foundations of the Statistical Approach in Mechanics*, Translated by Michael J. Moravcsik, Cornell U. Press, Ithaca, 1959.

Eringen, A. Cemal, *Nonlinear Theory of Continuous Media*, McGraw Hill, N.Y. 1962, note statements on p. 112 and 135.

Eyring, Henderson, Stover, and Eyring, *Statistical Mechanics and Dynamics*, Wiley, N.Y. 1964.

Feller, William, *An Introduction to Probability Theory and its Applications*, Wiley, N.Y. 1964.

Ferry, John D., *Viscoelastic Properties of Polymers*, Wiley, N.Y. 1961, 2nd ed. 1970.

Finck, Joseph Louis, *Thermodynamics from the Classic and Generalized Standpoints*, Bookman Associates, N.Y. 1955, revised and republished as *Topics in Physics*, 1962.

Flory, Paul J., *Principles of Polymer Chemistry*, Cornell University Press, Ithaca, N.Y., Chapters 11 and 12.

Flory, Paul J., *Statistical Mechanics of Chain Molecules*, Interscience, 1969.

Forgacs, O.L., Robertson, A.A., Mason, S.G., "The Hydrodynamic Behavior of Papermaking Fibers" in *Fundamentals of Papermaking Fibers*, Kenley, Surrey, England, pp. 447-473, 1958.

Forgacs, O.L., Mason, S.G., "Particle Motions in Sheared Suspensions, X Orbits of Flexible Threadlike Particles", *J. Col. Sci.*, **14**, 473-91, 1959.

Fowler, R.H., *Statistical Mechanics*, Cambridge, 1936.

Funatsu, Kazumor, and Mori, Yoshiro, "On the Viscoelastic Flow of Polymer Melts in the Nozzle and Reservoir", in *Proceedings of the Fifth International Congress on Rheology*, V. G. W. Harrison, ed. **4**, 537-50, University Park Press, Baltimore, (1970).

Garner, F.H., Nissan, A.H., Wood, G.F., "Thermodynamics and Rheological Behaviour of Elasto-Viscous Systems under Stress", *Royal Soc. London Phil. Trans.* **A243**, 37-66, 1950.

Garner, F.H., Nissan, A.H. *Nature*, **164**, 541, 1969.

Gauthier, F., Goldsmith, H.L., Mason, S.G., "The Kinetics of Flowing Dispersions, V Orientation of Distributions of Cylinders in Newtonian and Non-Newtonian Systems" *Kolloid-Z. u. A. Polymere*, **248**, 1000-15, 1971.

Gibbs, J. Willard, *Collected Works, Vol. 1 Thermodynamics*, Yale University Press, New Haven, 1928.

Glansdorff, P., Prigogine, I., *Thermodynamic Theory of Structure, Stability and Fluctuations*, Wiley, N.Y. 1971.

Good, R.J., "Thermal Effects in the Peeling Separation of an Adhesive Layer", in *Aspects of Adhesion*, D.J. Alner, ed., University of London Press, London, 1971.

Gosman, A.L., McCarty, R.D., Hunt, J.G., *Thermodynamic Properties of Argon from the Triple Point to 300 K at Pressures to 1000 Atmospheres*, National Bureau of Standards, National Standard Reference Data Series, National Bureau of Standards 27, 1969.

Goto, S., Kato, H., "Unstable Flow of Polymer Solutions" *Bull. J.S.M.E.*, **21**, (155) 854-60, 1978.

Guggenheim, E.A., *Thermodynamics*, 5th ed, Wiley, N.Y. 1967.

Haase, Rolf, *Thermodynamics of Irreversible Processes*, Addison-Wesley, Reading Mass. 1969.

Han, Chang Dae, "On Slit- and Capillary-Die Rheometry", *Trans, Soc. Rheol.* **18**, 163-90, 1974.

Han, Chang Dae, *Rheology of Polymer Processing*, Academic Press, N.Y. 1976.

Hayashida, K., Takahashi, J., Matusi, M., "Some Contributions to the Flow Behavior of Polymer Melts", *Proceedings of the Fifth International Congress of Theology*, Vol. 4, 537-50, University Park Press, Baltimore, 1970.

Hildebrand, Joel H., Scott, R.L., *The Solubility of Nonelectrolytes,* 3rd Ed. Reinhold, N.Y. 1950.

Hull, Harry H., "Viscoelasticity of Printing Ink", *American Inkmaker,* **29,** No. 9, 83-90, 1951, and in *Proceedings of the 3rd Annual Technical Meeting of the Graphic Arts,* May 7-9, 1951, pp. 83-90.

Hull, Harry H., "The Band Viscometer", *J. Coll. Sci.* **7,** 316-22, 1952.

Hull, Harry H., "The Normal Forces and Their Thermodynamic Significance", *Trans. Soc. Rheol.,* **5,** 115-31, 1961, and **13,** 167, 1969.

Hull, Harry H., "A Calculation for the Statistical Entropy Changes of Argon Gas over a Constant Energy Path", *Ind. Eng. Chem., Fundam.,* **12,** 257-8, 1973.

Hull, Harry H., "A Comparison of Two Laray Viscometers", *Research Department Report of Progress,* Graphic Arts Technical Foundation, Pittsburgh, PA, 235-42, 1971.

Hull, Harry H., "A Solution to the Time Average Problem of Statistical Mechanics", *Speculations in Science and Technology,* **3,** 41-48, 1980.

Hull, Harry H., unpublished data.

Isihara, A. *Statistical Physics,* Academic Press, N.Y. 1971.

Jost, W., *Diffusion,* Academic Press, N.Y. 1952.

Karnis, A., Mason, S.G., "Particle Motions in Sheared Suspensions, XIX Viscoelastic Media", *Trans. Soc. Rheol.,* **10,** 571-92, 1966.

Katchalsky, A., Curran, P.F., *Nonequilibrium Thermodynamics in Biophysics,* Harvard University Press, Cambridge, Mass., 1965.

Katz, Sydney M., "The thermodynamics of One-Component and Two-Component Elastic Systems" *Text. Res. J.,* **20,** *Text. Res. J.,* pp. 16-21 1950.

Kinchin, A.I., *Mathematical Foundations of Statistical Mechanics,* Translated by G. Gamow, Dover, N.Y. 1949.

Kirkwood, John G., "Statistical Mechanical Theory of Transport Processes, I General

Kotaka, T., Kurata, M., Tamara, M., "The Normal Stress Effects in Polymer Solutions", *J. Appl. Phys.* **30,** 1705-12, (1959).

Theory", *J. Chem. Phys.,* **14,** 180, 347, 1946.

Krigbaum, W.R., Roe, R., "Survey of the Theory of Rubberlike Elasticity", *Rubber Reviews,* **38,** 1039-68, 1965.

Landauer, Rolf, "Stability in the Dissipative Steady State", *Physics Today,* **31,** No. 11, 23-9, 1978.

Lavenda, Bernard H., *Thermodynamics of Irreversible Processes,* Wiley, N.Y., 1978.

Lewis, G.N., Randall, M., *Thermodynamics,* revised by Kenneth S. Pitzer and Leo Brewer, McGraw Hill, N.Y. 1961.

Li, J.C.M., Oriani, R.A., Darken, L.S., "The Thermodynamics of Stressed Solids", *Z. Phys. Chemie,* (NF) **49,** (Parts 3/4). 271-90, 1966.

Lindsay, Donald B., "Crystallographic Analysis", *The Crystal Front,* **2,** (No. 1), 6, June 1961.

Meixner, Joseph, "Entropy Concept in Nonequilibrium Thermodynamics", *J. Phys. Soc. Japan,* Suppl 1968, **26,** 212-4, 1969.

Michels, A., Wassenaar, T., Wolkers, G.J., Dawson, J., "Thermodynamic Properties of Xenon as a Function of Density up to 520 Amagat and as a Function of Pressure up to 2800 Atmospheres at Temperatures between 0°C and 150°C", *Physica,* **22,** 17-28, 1956.

Moore, Walter, J., *Physical Chemistry,* Prentice-Hall, Englewood Cliffs, N.J. 1972.

Mott, Sir Nevill, "Metal-Insulator Transitions", *Physics Today,* **31,** No. 11, 42-7, 1978.

Munster, Arnold, *Statistical Thermodynamics,* first English edition, Academic Press, N.Y. 1969.

Nielsen, Lawrence E., *Mechanical Properties of Polymers,* Reinhold, N.Y. 1962.

Nielsen, Lawrence E., *Polymer Rheology,* Marcel Dekker, N.Y. 1977.

Okagawa, A., Mason, S.G., "Kinetics of Flowing Dispersions, VII Oscillatory Behavior of Rods and Discs in Shear Flow", *J. Col. Sci.,* **45**, 330-58, 1973.

Okagawa, A., Mason, S.G., "Particle Motions in Sheared Suspensions, XXVII Configurations of Coiled Fibers in Shear Flows", *Can J. Chem.,* **53**, 2689-94, 1975.

Okubo, S., Hori, Y., "Experimental Determination of Secondary Normal Stress Difference in Annular Flow of Polymer Melts", *J. Soc. Rheol.,* **24**, 275-86, 1980.

Onsager, Lars, "Reciprocal Relations in Irreversible Processes", *Phys, Rev.,* **37**, I, 405, II2265, 1931.

Prigogine, I., Defay, R., *Chemical Thermodynamics,* translated and revised by D.H. Everett, Longmans Green, N.Y. 1954.

Prigogine, I., *Introduction to Thermodynamics of Irreversible Processes,* 2nd ed. Interscience, N.Y. 1955.

Prigogine I., "Time Structure, and Fluctuations", *Science, 201,* (No. 4385), 1, Sept., 1978.

Roseveare, W.E., Poore, L., "Applications of Thermodynamics to the Stretching of Cellulose Fibers", *Text. Res. J.,* **25**, pp. 709-14, 1955.

Racin, R., Bogue, D.C., "Molecular Weight Effects in Die Swell and in Shear Rheology", *J. Soc. Rheol.,* **23**, 263-80, 1979.

Roberts, J.E., "Pressure Distribution in Liquids in Laminar Shearing Motion and the Comparison with Predictions from Various Theories", in *Proceedings Second International Congress of Rheology,* V.G.W. Harrison, ed., Academic Press, New York, 1954.

Rosen, Milton J., *Surfactants and Interfacial Phenomena,* Wiley, 1978.

Rumscheidt, F.D., Mason, S.G., "Particle Motions in Sheared Suspensions, XII Deformation and Burst of Fluid Drops in Shear and Hyperbolic Flow", *J. Col. Sci.,* **16**, 238, 1961.

Rushbrooke, G.S., *Introduction to Statistical Mechanics,* Oxford University Press, Oxford, 1949.

Schreiber, H.P., Storey, S.H., Bagley, E.B., "Molecular Fractionation in the Flow of Polymeric Fluids", *Trans. Soc. Rheol.* **10**, p. 275-97 (1966).

Swinney, H.L., Gollub, J.P., "The Transition to Turbulence", *Phys. Today,* **31**, (No. 8), 41-9, Aug. 1978.

Tanner, R.I., "Some Methods for Estimating the Normal Stress Functions in Viscometric Flows", *Trans. Soc. Rheol.,* **14**, 483-507, 1970.

Taylor, G.I., "The Formation of Emulsions in Definable Fields of Flow", *Proc. Royal Soc.,* **A138**, 41, 1932, **A146**, 501, 1934.

Tine, T.W., Li, J.C.M., "Thermodynamics for Elastic Solids, General Formulation", *Phys. Rev.,* **106**, 1165-7, 1957.

Tollenaar, D., Bisschop, M.C., "The Bar Viscometer", *J. Col. Sci.,* **10**, 151-5, 1955.

Tolman, Richard C., *The Principles of Statistical Mechanics,* Oxford University Press, Oxford, 1938.

Rushbrooke, G.S., *Introduction to Statistical Mechanics,* Oxford University Press, Oxford, 1949.

Tordella, J.P., "Capillary Flow of Molten Polyethylene — A Photographic Study of Melt Fracture", *Trans. Soc. Rheol.,* **1**, 203-12, 1957.

Torza, S., Cox. R.G., Mason, S.G., "Particle Motions in Sheared Suspensions, XXVII Transient and Steady State Deformation and Burt of Liquid Drops", *J. Col. Interfac. Sci.,* **38**, 395-411, 1972.

Towle, L.C., Riecker, R.E., "Shear Strength of Grossly Deformed Solids", *Science,* **163**, 41-7, 1969.

Treloar, L.R.G., *The Physics of Rubber Elasticity,* 2nd ed., Oxford University Press, Oxford, 1958.

Treloar, L.R.G., "The Elasticity and Related Properties of Rubber". *Rep. of Progress in Phys.,* **36**, 755-826, 1973.

Tykodi, Ralph J., *Thermodynamics of the Steady State,* Macmillan, N.Y. 1967.

Tyrrel, H.J.V., *Diffusion and Heat Flow in Liquids,* Butterworths, London, 1961.

Wachholtz, Von F., Asbeck, W.K., "Ein Neus Verfahren zur Bestimmung der Viskositat bei hohen Schubspannungen und Schergeschwinding keiten", *Kolloid Zeitschrift,* **93,** (No. 3), 280-97, 1940, ibid **94,** (No. 1), 66-81, 1941.

Walters, Kenneth, *Rheometry,* Wiley, N.Y. 1975.

Ward, I.M., *Mechanical Properties of Solid Polymers.* Wiley-Interscience, N.Y. 1971.

Wriedt, H.A., Oriani, R.A., "Effect of Tensile and Compressive Elastic Stress on the Equilibrium Hydrogen Solubility in a Solid", *Acta Met.,* **18,** 753-60, 1970.

Yang, L., Horne, G.T., Pound, G.M., *Proceedings of a Symposium, Physical Metallurgy of Stress Corrosion Cracking, Pittsburgh, 1959,* Interscience, N.Y. 1959.

Zernike, J., *Chemical Phase Theory,* Uitgeversmaatschappij, AE, Ekluwer, Antwerp, 1955.

U.S. Patent No. 3,3321,908, assigned to Weizmann Institute by the inventors Aharon Katchalsky et.al.

Author Index

Adams, J., 10, 52
Asbeck, W.K., see Wachholtz, Von F.
Aubert, J.H., 127
Bagley, E.B., 114, see Schreiber, H.P.
Boltzman, L., VI
Brush, S.G., VI
Callen, H.B., 10
Casale, A., 52, 104
Chandrasekhar, S., 22
Cox, R. G., 100, 102
Denbigh, K.G., V, 8, 119, 113
Drigbaum, 52
Eyring, H., 22
Feller, W., 24
Ferry, J.D., Vi, 92
Flory, P.J., VI, 49
Forgacs, O.L., 100, 101
Funatsu, K., 96
Garner, F.H., 113, 114, 142
Gibbs, J.W., 49, 52
Glansdorff, P., V
Good, R.J., 60, 61, 62, 111
Gosman, A.L., 27-28
Goto, S., 115
Guggenheim, G.A., 8, 10, 41, 44, 80
Han, C.D., 115
Hull, H.H., 27, 93, 97, 111, 113, 115, 117, 141
Isihara, A., 22
Katchalsky, A., V, 71, 137
Katoka, T., 95
Kinchin, A.I., VI, 22
Kirkwood, J.G., 22
Kotaka, T., 95

Lavenda, B.H., V
Lindsay, D.B., 52
Lewis, G.N., V, 8, 27, 31, 32, 52, 68
Mason, S.G., 98, see Torza, Okagawa, Rumscheidt, Cox, and Forgacs
Meixner, J., 120
Michels, A., 29
Munster, A., 22
Nielsen, L.E., 52
Nissan, A.H., see Garner
Okagawa, A., 98, 100, 101, 103
Okubo, S., 117, 118
Onsager, L., V
Prigogine, I., V, 129
Roberts, J.E., 95
Rosen, M.J., 94
Rumscheidt, F.D., 96, 97, 98
Rushbrooke, G.S., 22
Schreiber, H.P., 126
Tanner, R.I., 115
Taylor, G.I., 96
Tollenaar, D., 117
Tolman, R.C., 22
Tordella, J.P., 114
Torza, S., 96, 97
Towle, L.C., 52
Treloar, L.R.G., 49, 50, 54, 70
Tykodi, R.J., 8
Wachholtz, Von F., 116
Walters, K., 94
Wood, G.F., see Garner
Wriedt, H.A., 71, 72
Zernike, J., 10

SUBJECT INDEX

Adiabatic, see paths adiabatic
Affinity, 129-136
Argon, 27-28, 59-60
Average,
 time, VI, 14, 18, 22, 26, 44
 arithmetic, 18, 141
 geometric, 18, 141
 root mean square, 18
Balance, 39-47, 121, 140
Barus effect, 96, 103, 113
Black box, V, 1-3
Capillaries, 123-125
Carnot cycle, 45, 81-84, 89, 104, 137
Centrifugal fields, 8
Chemical potential, 36-37, 89
Chemical reactions, 88, 129-138
Clapeyron equation, 32, 69
Compression, sign of, 35
Coupling of variables, V, 76, 123, 135-136
Dependent variables of state, 5-11, 49, 119, 129, 139
Die swell, 96
Drops, suspended in shear fields, 96-98
Elastic
 energy 46, 49-52
 deformation, 49-73, 91-118, 129-138
  non-rubberlike, 53, 71
  rubberlike, 54-65
Electric fields, 8, 10, 39
Enthalpy, 31-32
 extended, 32-33, 66
Entropy, VI, 21-30, 45, 66-69, 83-84, 107-109, 141
Equations of state, 5-11, 27, 49-50, 66, 73, 104, 119, 129, 139
Equilibrium, VII, 6, 39-46, 66, 140
 in an isolated system, 41
 at constant T and V, 42
 at constant P, T, 42
 at constant P, T and X, 44, 66
 in micro systems, 44
 in the presence of swelling, 69, 71
Extent of chemical reaction, 129-138
External work, 63, 65, 112
Filaments, suspended in shear fields, 100-103
Force, intermolecular, 26, 63, 93, 105
Free energy, 47, 50-53, 66, 86-89, 108-111, 126, 133, 135, 140
 non-additivity of, 88-89
Fugacity, 35
Galvanic cell, 52, 109, 132
Gas, ideal, 13-14, 58
 heavy monatomic, 16, 24-27
Gibbs function, 33-34, 42, 51, 105, 113, 121, 130
 extended, 33, 66, 121, 132

Gradients, 8, 72-73, 119-127
 concentration, 8, 115, 126
 pressure, 123-125
 shear, 115, 126
 temperature, 8, 121-125
Gravitational fields, 8, 39
Harmonic motion, 23, 55, 91
Heat of transport, 122-124
Helmholtz function, 33, 42
 extended, 66
Hook's law, 91
Hysteresis, 46-47, 104
Ideal gas, 13-14, 58
 rubber, 35, 53-55, 65
 mixing, 85-86
Independent variables of state, 5-11, 39-44, 66, 129, 139
Ink, printing, V, 93-95
Internal pressure, 17, 26, 59-60, 79
 relaxation, 76, 103
 shearing force, 107, 111
 tension, 60, 79, 107
 work, 17, 59-60, 65, 81, 85, 107, 112
Internal variables of state, 10, 46
Irreversibility, 55-56, 64, 85-86, 105
Kinetic energy, molecular, 26, 54, 79, 81, 93
Magnetic fields, 8, 10, 39
Maxwell model, 91-92
Mechanicochemical system, 137
Melt fracture, 115
Metastable state, 6, 40, 126
Mixing, 85-86
Models, 39-47, 139
 linear, 55, 91
 molecular, 24, 56, 100
 spring and dashpot, 55-56, 64, 80, 92-93
Moutier's theorem, 80
Newtonian fluids, 8, 92, 94
Non-balance, 46-47
Normal forces, 94-99
Osmotic pressure, 85-87
Palladium silver alloy, 71-72
Paths,
 adiabatic, 43-45, 51, 58-59, 81-84, 87
 components of, 28, 63-65, 75-77, 85-86
 constant energy, 27, 41, 87
 constant temperature, 81-84
 irreversible, see irreversibility
 pseudo-adiabatic, 81, 105, 126
 real, 45
 reversible, see reversibility
 steady state, 107
Phase rule, 9-10
Potential energy, intermolecular, 16, 26, 63-65, 79, 81, 93, 124

Pressure, 14,
  internal, 26, 60, 111
Probability, 22-24, 55, 106, 120
Relaxation, 46, 60, 80, 100
Restraints, 5-11, 40, 51, 87, 133, 140
Reversibility, 45, 55-56, 75-90, 105, 119
Rubber, 49-70
Separability of rev. and irrev., 75-90, 93, 105, 119
Shear, 10, 50-51, 96-104
Shear sensitive fluids, 8
Soret effect, 8, 125-126
Spring and dashpot models, 55-56, 63-64, 75-81, 86-87, 91-93
State,
  equations of, see equations of state
  properties of, V, 5-11, 140
Statistical mechanics, V, VI, 21
Steady state, V, 8, 40, 90, 91-118, 119-127
Stress (and strain), 7, 49, 106, 129-142
  effect on swelling, 69-71
  nonuniform, 72

Surface tension, 8
Swelling of polymers, 69-71
  of a metal, 71
Systems, 39-47
  isolated, 41
  at constant T and V, 42
  at constant T and P, 42
  micro, 44
Temperature, 15, 61-62, 88, 111-114, 142
Tension, 35, 50-51
Thermodynamic functions, 31-37, 66, 104
  definability of, 106, 108, 119-120
Thermostatics, 75
Velocity, molecular, 14, 24-26, 141
Viscoelasticity, 46, 55, 82, 91-118
Viscometers, 94, 114-118
Weissenberg effect, 94, 113,
  rheogoinometer, 93-94
Work, partitioning of 76-81
Xenon, 29
Zeroth law, 15

ADDENDUM

to

An Approach to Rheology Through Multivariable Thermodynamics

by

Harry H. Hull

This Addendum is dedicated to those who have read the book and are interested in extending its methods and concepts.

Sponsored by the Society of Plastics Engineers

Harry H. Hull
care of Deeds Associates
1300 Benedum-Trees Building
Pittsburgh, PA 15222

**Corrections to
ADDENDUM**

page,

A-2   Copyright by Harry H. Hull 1983

A-5   Figure A-5 should be turned up side down to make it consistent with the other figures.

A-10 On the last line the word "questions" should be "equations".

A-11 Second and third line — brackets should enclose ΔE^{PT} with $Cvy\ \Delta T$ to show their sum equals ΔE^{PS}, and brackets should enclose ΔE^{VT} with $Cvy\ \Delta T$ to show their sum equals ΔE^{VS}.

Addendum

to
An Approach to Rheology Through Multivariable Thermodynamics

Preface

After the publication of *An Approach to Rheology Through Multivariable Thermodynamics,* my thoughts on the relationship between rheology and thermodynamics continued to grow. They were further stimulated by comments received from readers. These thoughts have been collected in this *Addendum* which is now included with my book and is also available separately for those who previously purchased the book.

There have been two published reviews, one by Dr. John R. Collier in *Plastics Engineering,* **37**, No. 9, p. 53, (Sept. 81), and one by J.R.A. Pearson in *Journal of Fluid Mechanics,* **114**, p. 533, (Jan. 1982). I have also received a copy of a review to be published in *Polymer News.*

Professor Witold Brostow has commented on Chapter IV stating that he has also avoided the ergodic hypothesis and that his "approach to the same problem — in terms of information theory — seems to agree with and complement" mine. His approach is given in his book *Science of Materials* (1979).

I had hoped to receive more comments than I have; however, some of those which I have received have been very helpful. I hope this addendum stimulates further comments.

I again wish to thank the Society of Plastics Engineers and Dr. James L. Throne, Chairman of their Technical Volumes Committee, for their support in the publication of this book.

Harry H. Hull
September, 1982

The Elastic Free Energy and the Two Spring Thermodynamic Model
(Additions to Chapters VII and IX)

The maximum available work from a thermodynamic body is its free energy. When this work is available because of a body's elastic deformation, the actual value of the elastic free energy depends on the restraints on the system. (See Table VII-I.)

The thermodynamic model of a viscoelastic solid (Figure VII-10) and the thermodynamic model of a viscoelastic fluid (Figure IX-3) include two parallel springs along with a dashpot. One of these springs is labeled the molecular kinetic energy spring (or kinetic energy spring) and the other the intermolecular potential energy spring (or the potential energy spring). The properties that these two springs represent are very different, so it is desirable to have the symbols which represent these properties reflect these differences.

The concepts will be discussed first in terms of gases and then in terms of elastic solids and fluids. This is the method used in the book.

FIGURE A-1. The symbol for the molecular kinetic energy spring representing an ideal gas. Work done against this spring generates heat. Work done by this spring uses heat. The partial dashpot included represents the ability of the spring to relax without doing work but increasing its entropy. The arrow head indicates the one way direction of irreversible actions.

Figure A-1 shows the proposed symbol for the kinetic energy spring. Without other accompanying symbols it represents an ideal gas in which all the internal energy is in the kinetic energy of its molecules. It exerts a force equal and opposite to the external pressure and this force is proportional to the absolute temperature. It is convenient to designate the force exerted by this spring as Pk. If this gas (or spring) is held at constant temperature and allowed to expand doing reversible work against an external force this work is equal to,

$$-\int Pk \, dV = \int P \, dV = T \Delta S^T = Q \qquad (A-1)*$$

and the entropy of the gas is increased by

$$\Delta S^T = Q/T = \int (P/T) \, dV \qquad (A-2)*$$

The above is a reversible path.

The ideal gas can also expand into a vacuum doing no work, absorbing no heat, and without change in temperature. This is an irreversible path, but for the ideal gas the entropy change is the same as by the reversible path. (See Chapter IV.) Hence this spring has the property of being both reversible and irreversible, and the entropy change is the same over both paths.

The symbol in Figure A-1 includes both a spring to indicate its ability to use reversible paths and the arrow and dashpot to indicate its ability to use one way irreversible paths. When the irreversible path is used it becomes thermodynamically analogous to the dashpot.

FIGURE A-2. The symbol for the molecular kinetic energy spring representing an ideal rubber under tension.

Figure A-2 shows the symbol of the kinetic energy spring representing an ideal rubber under tension. The only difference from Figure A-1 is the direction of the arrows which indicate the direction of the external forces and the forces exerted by the body against them. The kinetic motion of the molecules exerts a pull equal and opposite to the pull placed on it by the external force X. It is convenient to designate the force generated by the kinetic motion of the molecules as Xk. In such an ideal rubber the molecules do not attract or repel each other just as in an ideal gas. The thermodynamic equations for an ideal rubber can be obtained by substituting X for P and Y for V and designating that V is held constant. For example, when reversible work is done, this is equal to,

$$\int Xk \, dY = -\int X \, dY = -T \, \Delta S^{VT} \tag{R-3}*$$

The change in entropy of the ideal rubber body over that reversible path is,

$$\Delta S^{VT} = Q/T = \int (X/T) \, dY \tag{R-4}*$$

The ideal rubber can also relax without doing work, or absorbing heat, or changing temperature — just as the ideal gas. The change in entropy is the same over this irreversible path as over the reversible path of equation R-4.

The intermolecular potential energy spring is very different from the kinetic energy spring; for all changes or paths for the spring itself are reversible — regardless of whether the other components of the system model are acting irreversible or reversibly. There is no means for this spring to relax. Because of this property the double spring was selected as its symbol. (See Figure A-3). The work done against this spring is the internal work.

FIGURE A-3. The symbol for the intermolecular potential energy spring. By itself it represents the ideal non-rubber. Work done by or against this spring is by or against the intermolecular forces, the sum of which is the internal pressure Pi or the analogous internal force Xi. Since its actions are always reversible, it is represented as a double spring.

The above two symbols are combined in parallel to represent the thermodynamic properties of gases and elastic bodies. Figure A-4 shows the two springs combined to represent the properties of real gases. Note that the kinetic energy spring is opposed by both the potential energy spring and the external pressure. The force per unit area exerted by the potential energy spring is called the internal pressure (Pi) and is equal to,

$$Pi = (\partial E/\partial V)_T = (T(\partial P/\partial T)_V - P) \qquad (A-5)$$

The force per unit area exerted by the kinetic energy spring (Pk) is equal to,

$$Pk = -(P + Pi) = -T(\partial P/\partial T)_V = -T(\partial S/\partial V)_T \qquad (A-6)$$

The maximum available work from a real gas at constant T is equal to the sum of the work available from the internal energy spring,

$$-Pi\, dV = \Delta E^T \qquad (A-7)$$

plus the work available from the kinetic energy spring

$$\int Pk\, dV = \int T\, \Delta S^T \qquad (A-8)$$

Since, $\qquad P + Pi + Pk = O \qquad (A-8a)$

this sums to,

$$\int P\, dV = \Delta E^T - T\, \Delta S^T = \Delta A^T \qquad (A-9)$$

which is the Helmholtz free energy of a gas.

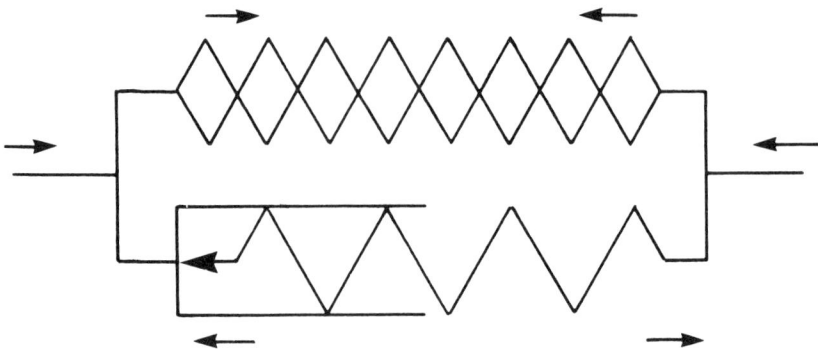

FIGURE A-4. The parallel kinetic energy spring and potential energy spring which represent the thermodynamic properties of a real gas.

The spring model in Figure A-4 also represents a gas under adiabatic or constant entropy conditions. In this case the sum of the internal and external work is converted to kinetic energy which remains in the gas raising its temperature on compression, or cooling the gas when it does work on expansion. Hence, we have the equation,

$$(P + Pi) \, dV = - Cv \, dT \tag{A-10}$$

or

$$Pi = - Cv \, (\partial T/\partial V)_s - P \tag{A-11}$$

It can be shown that equations (A-11) and (A-5) are thermodynamically equivalent, and hence the two definitions of Pi are equivalent.

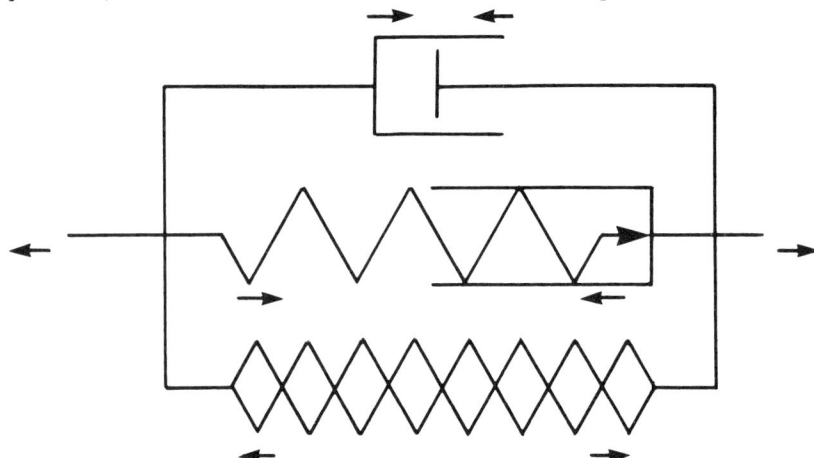

FIGURE A-5. The two parallel springs which are also parallel to a dashpot represent the thermodynamics of a viscoelastic solid.

Figure A-5 shows the use of these two springs as a model for a viscoelastic solid. In the following discussion it will be assumed that the experimental steps have been taken to obtain measurements of real bodies which eliminate the effects of hysteresis and viscous flow. In other words it is assumed that the effects of the dashpot have been eliminated and that the bodies are at equilibrium in terms of the three independent variables of state P, T, and X. (See page 43.)

If constant temperature and volume are assumed, the elastic free energy is the Helmholtz free energy,

$$\Delta A^{VT} = \Delta E^{VT} - T \Delta S^{VT} \qquad (A-12)$$

where ΔE^{VT} is the energy available from the potential energy spring and $- T \Delta S^{VT}$ is the energy available from the kinetic energy spring.

It is convenient to express the forces exerted by the springs as total forces rather than pressures per unit area. The force exerted by the potential energy spring is,

$$Xi = (\partial E/\partial Y)_{VT} = (T(\partial X/\partial T)_{VY} - X) \qquad (A-13)$$

and the force exerted by the kinetic energy spring is,

$$Xk = -(X + Xi) = -T(\partial X/\partial T)_{VY} = - T(\partial S/\partial Y)_{VT} \qquad (A-14)$$

which are equations analogous to those for a real gas (A-5) and (A-6).

This same spring model (Figure A-5) also represents a viscoelastic solid held at constant volume and constant entropy. In this case the elastic free energy is,

$$\Delta E^{VS} \qquad (A-15)$$

(See Table VIII-1.) In this case, just as with the adiabatic compression of a gas, the combined work done on the body by the combined internal and external forces is converted to heat or kinetic energy. We can write,

$$(X + Xi) \, dY = -Cvy \, dT \qquad (A-16)$$

or

$$Xi = - Cvy \, (\partial T/\partial Y)_{VS} - X \qquad (A-17)$$

Equations (A-15) and (A-13) are thermodynamically equivalent as definitions of Xi the internal force, i.e., the force exerted by the internal energy spring.

When the conditions are constant temperature and constant pressure the elastic free energy is a Gibbs free energy,

$$\Delta G^{PT} = \Delta E^{PT} - T \Delta S^{PT} + P \Delta V^{PT} \qquad (A-18)$$

where ΔE^{PT} is the energy available from the potential energy spring, and $- T \Delta S^{PT}$ is the energy available from the kinetic energy spring. It is necessary to add a third spring which corresponds to $P \Delta V^{PT}$. (See Figure A-6). Any change in volume also adds an internal work term. It now has two components,

$$\Delta E^{PT} = \int Xi' \, dY + \int Pi' \, dV \qquad (A-19)$$

where the primes designate their use in the above equation. The signs of these two internal work terms are generally opposite, but this will be discussed under thermoelasticity in the next section.

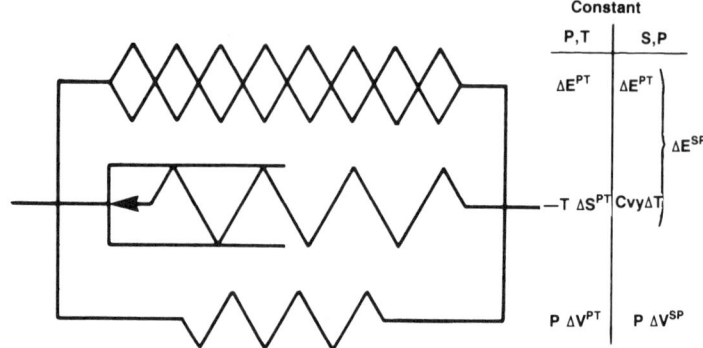

FIGURE A-6. The two springs in parallel with a $P\,dV$ spring which represent the thermodynamics of systems at constant pressure. The components of the equations for the free energies which the springs represent are listed at the right.

Since volume and the internal pressure Pi are functions of the extension Y (or the extensional force X), the internal work can be expressed in terms of a new variable Xi'' where,

$$\Delta E^{PT} = \int Xi''\,dY \tag{A-20}$$

For adiabatic conditions at constant pressure we have from the first law,

$$(X + Xi'')\,dY + P\,dV = Cvy\,dT \tag{A-21}$$

or

$$X + Xi'' = Cvy\,(\partial T/\partial Y)_{SP} - P\,(\partial V/\partial Y)_{SP} \tag{A-22}$$

The elastic free energy under adiabatic constant pressure conditions is,

$$\begin{aligned}\Delta H^{SP} &= \Delta E^{SP} + P\,\Delta V^{SP} \\ &= \Delta E^{TP} + Cvy\,\Delta T^{SP} + P\,\Delta V^{SP}\end{aligned} \tag{A-23}$$

The two free energies ΔG^{PT} and ΔH^{SP} are shown with their relationships to the springs in Figure A-6.

One may question whether there are elastic free energies other than those listed in Table VIII-1. There are conditions intermediate between those listed in that table. For example, a rubber band has available work even if the conditions listed in the table are not met. One can then write an intermediate free energy as,

$$\begin{aligned}\text{Work (max)} &= \Delta E^{?} - T\,\Delta S^{?} + P\,\Delta V^{?} \\ &= \Delta E^{VT} + \int^{?} Cvy\,dT - \int^{?} T\,dS + \int^{?} P\,dV\end{aligned} \tag{A-24}$$

where the question mark shows that these values must be set before the numerical value of the equation can be determined.

Table A-I gives a summary of the one to one correspondence between the components of the two-spring model and the components of the various free energy questions.

Table A-I. The relationship between the components of the two-spring model and the various elastic free energies.

Variables held constant	P, T	P, S	V, T	V, S
Potential energy spring	ΔE^{PT}	ΔE^{PT}	ΔE^{VT}	ΔE^{VT}
		ΔE^{PS}		ΔE^{VS}
Kinetic energy spring (reversible)	$-T \Delta S^{PT}_{rev.}$	$Cvy\, \Delta T$	$-T \Delta S^{VT}_{rev}$	$Cvy\, \Delta T$
$P\, dV$ spring	$P\, \Delta V^{PT}$	$P\, \Delta V^{PS}$	O	O
Free Energy (sum of the above)	ΔG^{PT}	ΔH^{PS}	ΔA^{VT}	ΔE^{VS}
Dashpot (irreversible work)	$T\, \Delta S_{ir.}$	$T\, dS_{ir.}$	$T\, \Delta S_{ir}$	$T\, dS_{ir}$
Location of irreversible S increase	heat sink	body	heat sink	body

Thermoelasticity

(A topic so old that it is omitted in most thermodynamic books)

Kelvin (W. Thomson) presented the theory of thermoelasticity in the 1850s. He also described it in the Ninth Edition of the Encyclopedia Britannica (1878).

The relationships can be developed as follows:
An equation of state is,

$$S = f(P, T, X) \tag{A-25}$$

Then from equation (II-11)

$$(\partial T/\partial X)_{SP} = -(\partial S/\partial X)_{TP}/(\partial S/\partial T)_{XP} \tag{A-26}$$

we have the Maxwell relationship (equation VII-64)

$$(\partial Y/\partial T)_{XP} = (\partial S/\partial X)_{TP} \tag{VII-64}$$

and the defining equation,

$$(\partial S/\partial T)_{XP} = Cpx/T \tag{A-27}$$

Combining the above three equations we have,

$$(\partial T/\partial X)_{XP} = -T(\partial Y/\partial T)_{XP}/Cpx = -\alpha T/Cpx \tag{A-28}$$

where α is $(\partial Y/\partial T)_{XP}$ or the coefficient of thermal expansion and Cpx is the specific heat at constant pressure and stress.

Most materials expand with increasing temperature so it follows from equation (A-28) that such bodies cool when placed under tension. This equation has been verified for many materials at room temperature and for two materials over a range of temperature (Rocca and Bever, 1950).

The two spring model (Figure A-4) contains the explanation for the cooling of a non-rubber thermodynamic body when placed under tension. The thermal motion of the molecules (i.e., the kinetic energy spring) helps the external forces do work against the intermolecular forces (the potential energy spring). When the thermal motion of the molecules does work this lowers the kinetic energy of the molecules thus lowering the temperature.

Materials which have a negative coefficient of thermal expansion behave in an opposite manner. Rubberlike materials *which are under tension* are of this type. The negative coefficent of thermal expansion is demonstrated by suspending a weight on a rubber body. When the temperature of the body is raised the weight will be lifted. Such bodies heat rather than cool when placed under tension in agreement with the equation.

When a rubberlike body is not under stress it has a positive thermal coefficient of expansion; so, when it is first placed under tension a slight cooling occurs. With further extensions the heating occurs. Treolar (p. 29) states that the initial cooling of rubber when stretched is caused by the increase in volume during the first ten percent of elongation. He states "at very low stresses the reduction of tension by thermal expansion exceeds the increase of tension to be expected from the kinetic theory of elasticity; the thermo-elastic inversion point is the elongation at which these two effects exactly balance." Staverman (1962) agrees that the initial cooling appears to be volume related.

Shear in rubberlike materials causes little or no change in volume; hence, there is no initial cooling with shear. Otherwise the thermodynamic behavior is similar to that under tension.

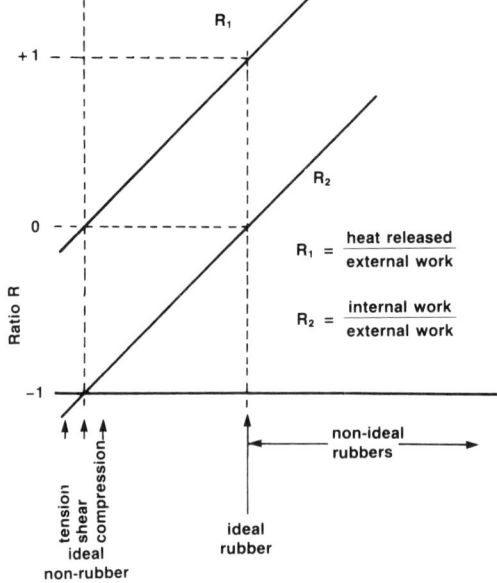

FIGURE A-7. A revision of Figure VII-11 to show the different thermal effects of tension, shear, and compression on non-rubbers.

The effects of thermoelastic heating and cooling of metals or non-rubbers is not shown properly in Figure VII-11. This chart has been revised to show the different effects of tension, shear, and compression (i.e., cooling, no effect, and heating) on non-rubbers. The revision is shown in Figure A-7.

The Change of the Gibbs Elastic Free Energy with Temperature
(An addition to Chapter XI)

The change of the Gibbs elastic free energy can be obtained by the same methods used for the Gibbs free energy of chemical reactions (Lewis and Randall, Prigogine). There are three relations.

From equation VII-43,

$$(\partial G/\partial T)_{PY} = -S \qquad \text{(VII-43)}$$

then

$$(\partial \Delta G^{PT}/\partial T)_{PY} = -\Delta S^{PT} \qquad \text{(A-29)}$$

By differentiation we also have,

$$(\partial(\Delta G^{PT}/T)/\partial T)_{PY} = (1/T)(\partial \Delta G^{PT}/\partial T)_{PY} - \Delta G^{PT}/T^2$$
$$= -(\Delta H^{PT}/T^2) \quad \text{(A-30)}$$

In a similar way

$$(\partial(\Delta G^{PT}/T)/\partial(1/T))_{PY} = \Delta H^{PT} \qquad \text{(A-31)}$$

Note that in order for these equations to be valid, Y (in this case the displacement or strain) must be held constant. Most texts do not note this; however, Prigogine (page 53) does the equivalent in applying these equations to chemical reactions. He shows that ξ (the extent of chemical reactions) is held constant, which is analogous to holding Y constant.

Many other equations from Prigogine may be converted for use in rheology by replacing the affinity $\mathbf{A}$ with X and ξ with Y.

The Internal Pressure of a Molten Polymer
(Some sample calculations)

Equations A-5 through A-11 were developed using a real gas as a model. However, liquids and solids could serve equally well as models for the development of these equations. We can apply equation A-11 to a molten polymer,

$$Pi = -Cv\ (\partial T/\partial V)_S - P \qquad \text{(A-11)}$$

Since Pi is much larger than atmospheric pressure for liquids and solids, this can be written,

$$Pi = -Cv\ (\partial T/\partial V)_S = -C\ (\partial P/\partial V)_S\ (\partial T/\partial P)_S \qquad \text{(A-32)†}$$

$$= (Cv\ Ms/V)\ (\partial T/\partial P)_S \qquad \text{(A-33)†}$$

where Ms is the adiabatic bulk modulus

$$Ms = -V\ (\partial P/\partial V)_S \qquad \text{(A-34)}$$

and † indicates that an approximation has been used in developing the equation.

The validity of equation A-33 may be checked by using the bulk modulus at constant temperature as an approximate for the bulk modulus at constant entropy. This was done for linear polyethylene at 190 °C using the following values,

		Source
$(\partial T/\partial P)_S$	1.5×10^{-7} °C/Nm^{-2}	Cogswell (1981)
Specific heat	2.5 $Kj/Kg\,°C$	Tsujita (1972)
Density	760 Kg/m^3	Cogswell (1981-A)
Bulk modulus	$0.9 \times 10^9\ N/m^2$	Tsujita (1972)

This gives,

$$Pi = 2.61 \times 10^8\ N/m^2$$

Tsujita, et al, (1972) have calculated the internal pressure of linear polyethylene using equation A-5,

$$Pi = T\ (\partial P/\partial T)_V - P \qquad \text{(A-5)}$$

The value read from a small graph in their paper and converted to the same units is,

$$Pi = 2.5 \times 10^8\ N/m^2$$

A-16

which is a reasonable agreement between the two methods.

Cogswell (1981) has studied the change in viscosity of molten polymers when placed under high pressure. He was concerned with the effects of high pressures on the flow of polymers during molding. He showed empirically there exists a straight line relationship between $(\partial T/\partial P)_S$ and $(\partial T/\partial P)_\eta$ where η is viscosity. Since,

$$\eta = f(P,T) \tag{A-35}$$

$$d\eta = (\partial \eta/\partial P)_T \, dP + (\partial \eta/\partial T)_P \, dT \tag{A-36}$$

then,

$$(\partial T/\partial P)_\eta = - (\partial \eta/\partial P)_T / (\partial \eta/\partial T)_P \tag{A-37}$$

Cogswell empirical relationship

$$(\partial T/\partial P)_S \cong c \, (\partial T/\partial P)_\eta \tag{A-38}$$

where c is a constant, can then be extended to,

$$Pi \cong c(\partial T/\partial P)_\eta (Cv \; Ms \; V)$$

$$\cong c(Cv \; Ms \; V) \, (\partial \eta/\partial P)_T / (\partial \eta/\partial T)_P \tag{A-39}$$

from equations (A-37) and (A-33).

Extensional Flow

(An addition for the end of Chapter IX)

The simplest example of extensional flow occurs during the uniform slimming down of a round solid body when placed under tension. (See Figure VII-2). Such flow occurs in specimens placed in tensile testing machines. Another example of extensional flow is the extension of extruded filaments which is done by pulling on the filament after extrusion. This is a common practice in the manufacture of synthetic fibers, for it gives these fibers more desirable physical properties. For a general discussion of extensional flow see Petrie (1979).

If the material undergoing extension flow is a Newtonian fluid, there is a direct relationship between the extension viscosity and the viscosity of that same fluid in shear. If the fluid is not Newtonian, there is no direct relationship between the flow in shear and extensional flow. It is reasonable to hypothesize that the different conformations taken by the molecules in the two flows explains the difference. In shear flow, the molecules roll into balls or cylinders (see Figures IX-9,10,11); while in extensional flow, the molecules are extended in the direction of flow. In concentrated solutions or melts, this behavior is modified by the entanglements between adjacent molecules.

However, with both types of flow the thermal motion of the molecules (i.e., the kinetic energy spring with a force of Xk) will do work to restore their average conformation to that of the unstrained state. This is the main source of rubberlike elasticity in both solids and fluids in shear and extensional flow and deformation. The intermolecular attractions (i.e., the potential energy spring with a force of Xi) is of the same sign as the external force X, so Xi tends to lessen the resistance to elastic deformation and make a "softer" rubber.

When thermodynamic bodies are in "equilibrium" in terms of the three independent variables of state (i.e., P, T, and X) the definitions of the free energies for shear and tension are identical. (See page 43.) Such conditions are attained with solid elastomers.

With the extensional flow such as occurs with the drawing of filaments there exists a steady state with accompanying gradients. (See Chapter X.) It is reasonable to assume that since the flow is very rapid the gradients in temperature and stress cause little or no gradients in composition or related effects. Hence, the gradients would contribute little to the elastic free energy and the correct thermodynamic equation for free energy would be selected from Table VII-1 according to the appropriate conditions.

With extensional flow there will probably be an appreciable amount of heat released. This will be equal to the sum of the external and internal work. (See page 63.)

Many elongational viscometers have been constructed which simulate the process of extruding and then drawing down filaments. Further measurements are needed — such as the elastic free energy and the amount of heat released. Such measurements are not simple, but the characterization of the process without them or related measurements is incomplete.

The above described flows are referred to as monoaxial. Biaxial flows also exist and are important in blow molding, vacuum forming, film blowing, and foaming processes. Biaxial viscometers have been described by Raible, et al. (1981). Biaxial flow can also be obtained by squeezing, and a viscometer which utilizes this method has been described. (Chatraei, 1981).

Probability in Periodic Motion
(Some thought on a concept in Chapter IV)

When equation IV-5 is supplied to simple harmonic motion the probability w that a point oscillating in simple harmonic motion will be in a section of its path dx is,

$$w = a\, dx/\pi \dot{x} \qquad (A\text{-}40)$$

where a is the amplitude and $\dot{x}$ is the velocity of the point at x.

The plot of probability versus position x is shown in Figure A-8. The probability is highest at the extremes of the motion and goes to infinity where the velocity passes through zero.

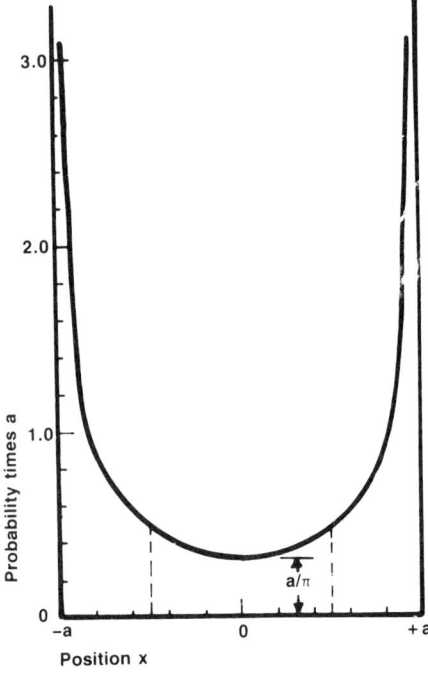

FIGURE A-8. A plot of the probability that a point in simple harmonic motion will be at a position x at a given random instant. The equation has been normalized so that the area under the curve is equal to the probability of one. However, the probability at the extremes goes to infinity.

Since the velocity must pass through two points with zero velocity in all periodic motions, all periodic motions have at least two positions at which the probability at that position is infinity. Any molecule or part of molecule which has a periodic motion and obeys the laws of Newtonian mechanics must have a probability distribution somewhat similar to that shown in Figure A-8, with maximums at the extremes of the motion.

The equation for the probability w that a point in simple harmonic motion will have the velocity in the range of $d\dot{x}$ can be obtained from equation IV-7,

$$w = a\, d\dot{x}/\pi\ddot{x} \tag{A-41}$$

where $\ddot{x}$ is the acceleration.

When this probability is plotted against velocity a graph identical to Figure A-8 is obtained, in which the probability becomes infinite at the extremes and has a minimum in the center. If the probability of a velocity is plotted against position x, the probability approaches infinity at the center and has a minimum at the extremes.

I have not found these relationships in the literature; however, they are so obvious that they must be there. These graphs do help visualize the motion of the molecules in terms of Newtonian mechanics.

Melt Flow Instability
(Reference page 115)

J.R.A. Pearson states in his review of this book (1982) "nor is it unambiguously clear how melt flow instability can be explained in terms of Gibbs free energy (p. 115)".

Instead of *Gibbs free energy* the explanation should have been in terms of *elastic free energy*. The specific constraints involved for the conditions which produce that melt flow instability are not defined. They are probably not constant P and T as they would be if the Gibbs free energy were involved; the constraints could be at any of the levels listed in Table VII-I or at levels between those listed levels. (See equation A-24.)

It is a general principle of thermodynamics that processes which lower the free energy are thermodynamically feasible — though they do not necessarily proceed. A fluid undergoing shear, which contains an elastic free energy, can lower its free energy by flowing to an area which is not undergoing shear, where it can relax and release its free energy. When it flows out of the shear field it will push the material it replaces into the shear field where an elastic free energy is imposed on it. Both the relaxation and the imposition of free energy are processes which take finite time, so a periodic exchange of material between the shear field and the non-shear field is quite likely. This then causes the melt flow instability.

It follows from the above reasoning that, whenever there is a shear gradient in a viscoelastic fluid (and all other variables are held constant) a free energy gradient is generated which can be a source of instability. Such shear gradients exist in the flow in pipes or capillaries. However, flow in pipes or capillaries is stable at moderate rates of flow. It follows that such flows must usually be considered to be in a metastable state. (See paragraph f, page 6.)

All flow in pipes or capillaries is not necessarily in the metastable state, for when such flows are allowed to continue concentration gradients can form which can lessen or eliminate the free energy gradient. (See Chapter X.)

Some Additional Comments

In chemical thermodynamics the Gibbs free energy is used almost exclusively. Undoubtedly one reason for this is the practice of working with dilute solutions and extrapolating to infinite dilution. Where the solvent is predominant it acts as a heat sink for the measured component. This causes the temperature to be effectively constant. Since measurements are also made at atmospheric pressure the free energy is a Gibbs free energy.

In this Addendum the emphasis has been that the free energy depends on the restrictions on the system. We are not restricted to constant temperature and pressure systems. It is my opinion that measurements under adiabatic conditions have not been properly exploited by the rheologists. However the measurements needed will require different instruments — which include a means for measuring rapid small temperature changes, and a means for the rapid imposition and release of forces. Figures VII-8 and IX-13 show the results of simple experiments of this type. However, the amount of temperature rise in these experiments is much higher than would be expected for many experiments, so the measurement of small changes in temperature accurately and quickly are needed. The work of Fuller, et al, (1975) is an example of what can be done. He used both temperature sensitive liquid crystals and an infrared detector to measure temperature changes at the tip of fast moving cracks in glassy polymers.

References

Brostow, W., *Science of Materials*, Wiley, New York, N.Y. 1979.

Chatraei, Sh., and Macosko, C. W., "Lubricated Squeezing Flow: A New Biaxial Extensional Rheometer", *J. Rheo.*, **25**, 433-43, 1981.

Cogswell, F.N., *Polymer Melt Rheology*, Wiley, New York, N.Y. 1981.

Cogswell, F.N. Personal communication, 1981-A.

Fuller, K.N.G., Fox, P.G., Field, J.E., "The Temperature Rise at the Tip of Fast-Moving Cracks in Glossy Polymers", *Proc. R. Soc. Lond.* **A**, **341**, 537-557, 1975.

Petrie, C.J.S., *Elongational Flows*, Pitman, London, 1979.

Munstedt, H., Middleman, S., "A comparison of Elongational Rheology as Measured in the Universal Extension Rheometer and by the Bubble Collapse Method", *J. Rheo.*, **25**, 29-40, 1981.

Raible, J., Meissner, T. Stephenson, S.E., "Rotary Clamp in Uniaxial and Biaxial Extensional Rheometry of Polymer Melts", *J. Rheo.*, **25**, 1-28, 1981.

Rocca, R., Bever, M.B. "The Thermoelastic Effect in Iron and Nickel as a Function of Temperature", *Trans. Met. Soc. AIME*, **188**, 328-33, 1950.

Staverman, A.J., "Thermodynamics of Polymers" in *Handbook der Physik*, edited by S. Flugge, V. XIII, Springer Verlag, Berlin, 1962.

Tsujita, Y., Nose, T., Hata, T., "Thermodynamic Properties of Polyethylene and Eicosane", *Polymer J.*, **3**, 581-6, 1972.

Zenner, Clarence, *Elasticity and Anelasticity of Metals*, University of Chicago Press, Chicago, 1948.

Corrections for "An Approach to Rheology Through Multivariable Thermodynamics"

p. 21   In the last sentence in the first paragraph insert "on the average" so the sentence reads " — the entropy of an isolated system can only on the average increase or remain the same."

p. 22   Change "Boltzman (1964)" to "Boltzmann (English translation in 1964)".

p. 25   At the end of the paragraph containing equation (12) change "in line of path" to "tangent to the path" and add an explanatory sentence so it reads: " — f is an average force tangent to the path acting on that molecule (within that $\triangle \dot{x}$). It is necessary that f be the tangential component of the total force acting on that molecule for only that component changes the velocity of the molecule and hence does work. The component normal to the path changes the direction of the path but does no work."

p. 25   Change equation (13) and the following phrase to read,

$$W = (w_c)^N = k(1/\bar{f}')^N \quad (13)$$

"where $\bar{f}'$ is the geometric mean of f.†"

p. 25   Change the word "average" to "mean" in the line after equation (15) so that it reads, "where $\bar{f}_1'$ and $\bar{f}_2'$ are the mean intermolecular forces —".

p. 25   Change the footnote at the bottom of the page to read "†See note in Chapter XII, p. 141."

p. 26   Change equation (18) to read,

$$E = E_k + E_p \quad (18)^{**}$$

p. 26   In the paragraph following equation (19), change "in line of" to "tangent to" so the sentence reads, "— that component of the intermolecular forces which act tangent to the molecular paths —".

p. 27   Place the partial derivative sign in equation (22) so the first part of the equation reads,

$$P_iv = (\partial E/\partial V)_T v =$$

p. 28   Table I. Change the number in column 4 in the next to the last row from 19188 to 18830.

p. 29   Table III. Change the heading of column 4 to $\Delta s''$ from $\Delta v''$.

p. 61   The first sentence at the top of the page should read, "A very small thermocouple (0.001 inch wires) was placed in the adhesive between two tapes and then the tapes were peeled apart with an Instron."

p. 65   Figure 11. The plots of R_1 and R_2 should extend to the left of the ordinate as metals cool when placed under tension. This is thermoelastic cooling and was extensively discussed by Lord Kelvin.

p. 149   Change "Boltzman" to "Boltzmann".

p. 155   Change "Boltzman" to "Boltzmann" and change "G.A. Guggenheim" to "E.A. Guggenheim".

(Please notify author of any additional errors found.)